A Guidebook to Earthquake Scientific Investigation

Institute of Earthquake Forecasting, China
Earthquake Administration

A Guidebook to Earthquake Scientific Investigation

地震出版社
Seismological Press

Springer

Institute of Earthquake Forecasting, China
Earthquake Administration
Beijing, China

ISBN 978-981-96-0420-3 ISBN 978-981-96-0418-0 (eBook)
https://doi.org/10.1007/978-981-96-0418-0

Jointly published with Seismological Press
The print edition is not for sale in China (Mainland). Customers from China (Mainland) please order the print book from: Seismological Press.
ISBN of the Co-Publisher's edition: 978-7-5028-5653-3

Editor in Charge: Guijuan Guo
Proofreader: Ying Ling

This Springer imprint is published by the registered company Springer Nature Singapore Pte Ltd.
The registered company address is: 152 Beach Road, #21-01/04 Gateway East, Singapore 189721, Singapore

Members of the Compilation Group

Steering Group

Shi Che, Weixin Zhou, Feng Wang, Xiaodong Zhang, Zhongliang Wu, Ying Li, Zhigang Shao, Xinjian Shan, Honglin He, Zhifeng Ding, Baitao Sun, Qinjian Tian, Jian Wu, Tao Chen, Longquan Zhou, Gan Ma, Ming Ma

Director

Guojie Meng

Managing Editors

Renwei Xiong, Wenqiao Li, Aixia Dou, Xiaoqing Wang, Lin Ding, Guojie Meng, Wenxin Wei, Shunying Hong, Guoqiang Zhao, Weiwei Wu, Xiaocheng Zhou, Zhi Chen, Cuiping Zhao, Xun Wang, Shuaijun Wang, Chongyang Shen, Guangliang Yang, Hongbo Tan, Minzhang Hu, Hongtao Hao, Tengfa Cui, Yan Zhan, Jifa Ji, Weijun Wang, Jun Wang, Bin Chen, Ruizhi Wen, Yongxian Zhang, Qincai Wang, Hongwei Tu, Shengfeng Zhang, Feng Li, Weiwei Zhan

Editing Group

Renwei Xiong, Wenqiao Li, Guojie Meng

Preface

Large earthquake may cause socio-economic losses, while it also provides a large number of observational data for earthquake research which are difficult to obtain during interseismic periods. However, most co-seismic and early post-seismic information perish quickly in a short period after the occurrence of an earthquake. Therefore, the rapid initiation of comprehensive scientific investigation after an earthquake is a necessary prerequisite for "anatomizing the earthquake" and analyzing the disaster-causing factors, and an important way to advance earthquake research and earthquake disaster prevention and mitigation science and technology.

For a long time, China Earthquake Administration attaches great importance to the scientific investigation of earthquakes. The earthquake disaster prevention and mitigation work in China has been made rapid progress in major earthquakes; the history of earthquake science and technology progress is also a history of seismic scientific investigation. *Law of the People's Republic of China on Protecting Against and Mitigating Earthquake Disasters* stipulates that relevant agencies of the State Council and governments should "organize the investigation and evaluation of earthquake disaster losses timely to provide a basis for earthquake emergency rescue, post-seismic transitional resettlement, recovery and reconstruction" and "carry out the investigation and evaluation of the damage mechanism of seismic activity on construction projects, to provide a scientific basis for revising and improving the mandatory standards for construction projects and for taking seismic reinforcement measures".

For more than half a century, China has carried out a wealth of seismic scientific investigations, and more than 200 case studies of $M \geq 5.0$ earthquakes have been summarized. Especially, the analyses and researches on the nine major earthquakes in the 1960s and 1970s have brought an excellent reputation domestically and internationally. After the 2001 Kunlun (M_S 8.1) earthquake and the 2008 Wenchuan (M_S 8.0) earthquake, China Earthquake Administration organized large-scale systematic scientific investigations into the two earthquakes. *Regulations on Post-Wenchuan Earthquake Rehabilitation and Reconstruction* stated clearly that the investigation and evaluation of earthquake disaster require that "earthquake agencies and earthquake monitoring network should collect and archive all data and information before, during and after the earthquake, and establish complete archives". More than 1200 researchers participated the scientific investigation into the Wenchuan earthquake, acquiring enormous firsthand data and setting up scientific investigation database. Based on the subsequent studies, new understandings are acquired on the mechanism of earthquake and its disaster, which provide scientific basis for the planning of post-Wenchuan earthquake rehabilitation and reconstruction. Scientific investigations have been all organized after the 2010 Yushu (*M* 7.1) earthquake, 2013 Lushan (*M* 7.0) earthquake, 2014 Ludian (*M* 6.5) earthquake, 2015 Pishan (*M* 6.5) earthquake, 2017 Jinghe (*M* 6.6) earthquake, 2017 Jiuzhaigou (*M* 7.0) earthquake, 2019 Changning (*M* 6.0) earthquake, 2021 Yangbi (*M* 6.4) earthquake, Maduo (*M* 7.4) earthquake and Luxian (*M* 6.0) earthquake, 2022 Menyuan (*M* 6.9) earthquake and other destructive earthquakes. Although the on-site scientific investigations have accumulated a wealth of practical experiences, some problems and shortcomings have been emerging during the investigation. For example, data accumulation before these earthquakes is not sufficient, the analyses are not deep enough and the time is limited in finishing the scientific earthquake investigations, the lack of detailed fundamental background data of earthquake zone causes the scientific investigation target too broad, leading to too many temporary basic observation after the earthquakes, and too long implementation period for scientific investigation. For example, the scientific investigation of the 2008 Wenchuan earthquake has 20 special topics, and the emergency scientific investigation and systematic scientific investigation lasted for more than six months. With the improvement of seismic monitoring capability, the on-site scientific investigation of earthquakes has been able to be developed to attain the standard of "rapid deployment and simple logistics", focusing on exploration of seismogenic structure and evaluation of the recurrence patterns of strong earthquakes, investigation and evaluation of engineering failure mechanism, review and evaluation of regional ground motion, effectiveness evaluation of regional seismic monitoring network, earthquake precursor recognition and forecasting methods testing and evaluation, regional earthquake disaster prevention and mitigation management system and management capability assessment, etc., which can greatly reduce the workload of scientific investigations. Even if the on-site working conditions are not available for a short time after the earthquake, we can use the existing data to carry out "virtual scientific investigation", so as to achieve the rapid response and outcome after earthquakes. It provides scientific and technological support for "anatomizing" and study earthquakes, understanding the earthquake preparation process, and preventing and mitigating earthquake disaster.

Scientific and technological innovation support on *Pilot Action Plan for Modernization of Earthquake Disaster Prevention and Mitigating in the New Era (2020–2022)* issued by China Earthquake Administration requires experts from various disciplines with scientific investigation experiences to compile *A Guidebook to Earthquake Scientific Investigation* (hereinafter, *the Guidebook*). *The Guidebook* aims to summarize and adopt the experience of previous seismic scientific investigations, form normative contents, procedures and approaches of scientific investigation work, promote the management, archival, sharing of work results and the standardization of scientific investigation, and better serve the earthquake disaster prevention and mitigation. The core content of *the Guidebook* is divided into 15 topics: seismogeology investigation, seismic disaster remote sensing comparative investigation and acquisition, crustal deformation observation, geochemical observation, source rupture process and seismic tectonic mechanism research, artificial seismic sounding, gravity detection, mobile gravity observation, magnetotelluric sounding, deep seismic reflection sounding, mobile seismologic station array observation, mobile geomagnetic observation, strong ground motion and engineering earthquake damage investigation, seismic sequence analysis, data collection and sharing services.

The Guidebook refers to a large number of earthquake-related national standards and industrial technical specifications and previous seismic schemes and achievements of scientific research, summarizes the experience and deficiency of previous seismic scientific investigations. On the basis of comprehensive consideration of the scientific characteristics and operability of *the Guidebook*, full discussion and extensive consultation are conducted during the compilation of the Guidebook. *The Guidebook* can provide important guidance and reference for the development of seismic scientific investigation and will continue to develop and improve aside with the practices of future scientific investigation.

The compilation of *the Guidebook* was under the auspices of the Institute of Earthquake Forecasting of China Earthquake Administration, jointly completed by the CEA-affiliated institutions: Institute of Geophysics, Institute of Geology, Institute of Engineering Mechanics, China Earthquake Disaster Prevention Center, Geophysical Exploration Center, and Hubei Earthquake Agency.

Principal researcher responsible for compilation: Guojie Meng.

The Preface and Chaps. 1 and 2 are mainly written by Guojie Meng, Renwei Xiong and Wenqiao Li.

The main compilers are as follows:

Chapter 3: Seismogeology Investigation by Qinjian Tian, Renwei Xiong, Wenqiao Li; Chap. 4: Seismic Disaster Remote Sensing Comparative Investigation and Acquisition by Aixia Dou, Xiaoqing Wang, Ling Ding; Chap. 5: Crustal Deformation Observation by Guojie Meng, Wenxin Wei, Shunying Hong, Guoqiang Zhao, Weiwei Wu; Chap. 6: Geochemical Observation by Xiaocheng Zhou, Zhi Chen; Chap. 7: Source Rupture Process and Seismic Tectonic Mechanism Research by Cuiping Zhao, Xun Wang; Chap. 8: Artificial Seismic Sounding by Shuaijun Wang; Chap. 9: Gravity Detection by Chongyang Shen, Guangliang Yang, Hongbo Tan; Chap. 10: Mobile Gravity Observation by Chongyang Shen, Minzhang Hu, Hongtao Hao; Chap. 11: Magnetotelluric Sounding by Tengfa Cui, Yan Zhan; Chap. 12: Deep Seismic Reflection Sounding by Jifa Ji; Chap. 13: Mobile Seismologic Array Observation by Weijun Wang, Jun Wang; Chap. 14: Mobile Geomagnetic Observation by Bin Chen; Chap. 15: Strong Ground Motion and Engineering Earthquake Damage Investigation by Ruizhi Wen; Chap. 16: Seismic Sequence Analysis by Yongxian Zhang, Qincai Wang, Hongwei TU, Shengfeng Zhang; Chap. 17: Data Collection and Sharing Services by Feng Li, Weiwei Zhan, Wenqiao Li, Renwei Xiong.

The compilation of *the Guidebook* has received support from leaders of China Earthquake Administration and its departments. The leaders and experts, Shi Che,

Weixin Zhou, Feng Wang, Xiaodong Zhang, Zhongliang Wu, Ying Li, Zhigang Shao, Xinjian Shan, Honglin He, Zhifeng Ding, Botao Sun, Qinjian Tian, Jian Wu, Tao Chen, Longquan Zhou, Gan Ma and Ming Ma have provided support and assistance during the compilation of *the Guidebook*.

Jinwei Ren, Xuejun Qiao, Zhikun Ren, Lijun Chang, Youjin Su, Feng Long put forward valuable comments on the first draft of *the Guidebook*. Weijun Gan, Yonghong Duan, Changsheng Jiang, Xuchuan Lin, Xiaolong Sun, Ji Tang, Jian Wu, Yan Xue, Daoyang Yuan, Yiqing Zhu reviewed the submitted manuscript in detail and put forward very constructive suggestions. We would like to express our thanks to Peng Li, Long Wang and Chaozhong Hu, etc. from the Institute of Earthquake Forecasting of the China Earthquake Administration for providing much support in organization and coordination.

The publication of *the Guidebook* is financially supported by the National Key Research and Development Program (2019 YFE0108900) and the Pilot Project of Scientific and Technological Innovation Modernization of the Institute of Earthquake Forecasting of China Earthquake Administration (2020 IEF1204).

Beijing, China Guojie Meng

Contents

Chapter 1
Overview of the Compilation of a Guidebook to Earthquake Scientific Investigation

1.1 Compilation Objectives

The compilation objectives of *A Guidebook to Earthquake Scientific Investigation* (hereinafter, *the Guidebook*) are to summarize and take the experience of previous seismic scientific investigations, form normative investigation contents, procedures, methods and the results submitting scheme, etc., which promotes the deepening of "earthquake anatomizing" work, the operation and standardization of seismic investigation and serves better the earthquake disaster prevention and mitigation.

1.2 Main Contents

After the occurrence of a major earthquake, we should track rapidly the development process of its sequence and determine the properties of the sequence and its influence on the trend of surrounding potential large earthquakes. In the following seismic investigation, a number of thematic researches are often carried out: the basic seismogenic units and structures, focal rupture process and seismogenic mechanism, earthquake propagation and disaster-causing mechanism, etc. On the basis of a large amount of data collection, full discussion and extensive consultation with experts, we summarize the experience and shortcomings of previous seismic investigation, and provide a set of scientific and operational seismic investigation guidelines for rapid use after a large earthquake. Its core contents include the following 15 topics: seismogeology investigation, seismic disaster remote sensing comparative investigation and acquisition, crustal deformation observation, geochemical observation, source rupture process and seismic tectonic mechanism research, artificial seismic sounding, gravity detection, mobile gravity observation, magnetotelluric sounding, deep seismic reflection sounding, mobile seismologic array observation, mobile geomagnetic observation, strong ground motion and

Institute of Earthquake Forecasting, China Earthquake Administration,
A Guidebook to Earthquake Scientific Investigation,
https://doi.org/10.1007/978-981-96-0418-0_1

engineering earthquake damage investigation, seismic sequence determination, data collection and sharing services.

1.3 Instructions

Seismic investigation should meet the new requirements of earthquake disaster prevention and mitigation in the new era, which is necessary for social economic development and the people's aspiration for a better life in China. The seismic investigation is oriented to the modernization of earthquake disaster prevention and mitigation, which is very important for understanding and summarizing the experience of earthquake disaster prevention and mitigation. It needs comprehensive research of multi-phase and multi-scale, as well as the collaborative participation of multi-disciplines and multi-means. On the basis of summarizing previous seismic investigation, combined with the current development of industry and social needs of scientific investigations, the Guidebook covers as comprehensively as possible various work contents and conventional technical means involved in seismic investigations. In recent years, government and society have paid more attention to the timeliness of seismic investigations and the rapidity of the output of results. The boundary between post-seismic emergency scientific investigations and systematic scientific investigations has been gradually broken, which puts forward higher requirements for the organization and undertaking of seismic investigations.

Generally speaking, after the occurrence of a destructive earthquake, in order to meet the needs of post-seismic emergency work, the seismic investigation is actually an important part of the seismic field work and is more often carried out in the form of post-seismic emergency investigation. After an earthquake, in addition to meet the demand of emergency, a comprehensive understanding of earthquake preparation, occurrence mechanism, evolution process and evaluating the earthquake disaster and its subsequent impacts provide scientific basis for the post-seismic relief work, government decision-making and post-seismic reconstruction and make technology support for judgment of earthquake forecasting after earthquake, seismic disaster prevention and a series of earthquake disasters mitigating work. Therefore, on the basis of the emergency scientific investigation, it is necessary to organize the comprehensive scientific investigation that lasts longer (months or even years) and invests more human and material resources after earthquake.

As an important part of earthquake disaster prevention and mitigation, seismic investigation should explore the transition from "post-disaster" to "predisaster". From this perspective, the time and space span and work scale of seismic investigation should be extended greatly. As a more generalized seismic investigation, its content can be divided into two parts. One part is usually carried out after earthquake, such as aftershock characteristics research, surface rupture investigation, seismogenic structure determination, seismic rupture process inversion, co-seismic and post-seismic geophysical and geochemical observation, engineering earthquake damage investigation, earthquake disaster investigation and loss assessment, etc. The other part can be carried out before earthquake, such as investigation into deep seismogenic environment, intensive observation of geophysics and geochemistry, research of long-term activity behavior of potential seismogenic faults and the comprehensive observation and analysis of precursor evolution.

The scientific problems of preseismic scientific investigation are often more complex, and degree of uncertainty for them is higher, which requires a certain degree of foresight. At present, there is not much practical experience to refer to. It is suggested that we could combine the work of "earthquake potential zone" and "key zones for earthquake monitoring and prevention" to plan, layout and implement preseismic investigation in advance. *The Guidebook* is mainly compiled for the scientific investigation after earthquake, and there is no systematic discussion on the scientific investigation of enhanced monitoring before earthquakes. However, if the location of the potential source and the possible rupture characteristics of the seismogenic fault can be roughly determined before earthquake occurrence, according to scheme of "seismic scenario investigation" and "inverse investigation", in view of the potential strong earthquakes occur scene model, draw lessons from "*the Guidebook*" in the seismic investigation scientific questions, research contents, the thinking of work, methods, techniques and related technical specification, the relevant investigations in potential earthquake source area could be carried out in advance, such as seismogenic environment and evolution of earthquake preparation. For specific investigation objectives, investigation methods, contents and expected outcomes, please refer to the detailed contents of corresponding modules in Tables 1.1, 1.2, 1.3, 1.4 and 1.5. With persistent strengthening of preseismic intensive scientific investigation practice, the experience gained will surely "feedback" *the Guidebook* in the future, enriching *the Guidebook*, providing valuable experience for further improvement.

The previous practice of major seismic investigation shows that the seismic investigation implementation should be rapid and comprehensive. It is necessary to carry out field work timely after the earthquake. Therefore, making a scientific investigation plan rapidly after large earthquake is an essential requirement for ensuring success of the seismic investigation. Usually, seismic investigation committee will hold a timely symposium short after large earthquake and put forward scientific problems that needed to be solved, by considering the location, magnitude of the earthquake and degree of investigation before the earthquake, and selecting targeted investigation means. *The Guidebook* can play a useful reference to rapidly form a scientific investigation plan.

According to the modular work content of *the Guidebook*, seismic investigation involves seismogenic structure investigation (Table 1.1), seismogenic mechanism investigation (Table 1.2), earthquake evolution process investigation (Table 1.3), influence of the studied earthquake on regional earthquake hazard analysis (Table 1.4) and earthquake disaster mechanism investigation (Table 1.5). For any specific investigation content, it is not necessary to take all the investigation means. According to the geographical surroundings of the epicentral area, geomorphic features, transportation conditions, cultural conditions, the characteristics of the earthquake in its early period and the degree of investigation before the earthquake, we can select the necessary means of investigation from the above five modules and form the

Table 1.1 Seismogenic structure investigation

Sequence number	Investigation objectives	Investigation approach	Contents	Expected outcomes
1	Determine the location and geometric characteristics of seismogenic structure, provide a basis for researching the source characteristics, rupture process and disaster distribution of earthquakes	Seismogeology investigation	Using remote sensing interpretation and field investigation, carry out active structures, map surface rupture zones of surface rupture earthquakes and investigate seismogenic structural environment and seismic geological disasters comprehensively	Characteristics of seismic surface rupture and displacement distribution, classification and distribution of seismic geological hazards, seismogenic structure geometry, kinematic characteristics and long-term activity behavior
2		Research on precise seismic location	Using different seismic positioning methods, obtain precise seismic catalog, comparative analyses to the seismic events	Geometry of seismogenic structure and migration characteristics of aftershocks
3		Research on co-seismic deformation field	Using GNSS data and InSAR image processing, obtain the co-seismic deformation field and invert for co-seismic rupture model	Source geometric model and slip distribution characteristics

(continued)

Table 1.1 (continued)

Sequence number	Investigation objectives	Investigation approach	Contents	Expected outcomes
4		Research on source rupture process	Using seismic waveform processing and analysis, calculate focal mechanism solutions of main shocks and large aftershocks, carry out inversion for seismogenic structure analysis and spatio-temporal source rupture process	Geometry of seismogenic structure, and spatial-temporal distribution of source rupture history
5		Comprehensive interpretation of seismogenic structures	Obtain overall scientific understanding of seismogenic structure by comprehensive analyses of the above four aspects of investigations, analyze the shortcomings of preearthquake monitoring and scientific investigation	Comprehensive interpretation of seismogenic structures and suggestions for further study and improvement of earthquake preparedness

corresponding investigation team, by considering the geographic distribution of the professional earthquake investigators.

According to *Earthquake Emergency Response Service Grading (Provisional), China Earthquake Administration, and List of Earthquake Emergency Response Services over 12 hours post-seismic, China Earthquake Administration*, it is necessary to rapidly initiate scientific investigation. The first step of scientific investigation is to formulate scientific investigation plan. How to formulate scientific and reasonable investigation plan quickly and efficiently is the key to quality of scientific investigation. *The Guidebook* can provide preliminary ideas and references for the formulation of the scientific investigation plan and provide guidance for the specific work in the implementation process of the scientific investigation plan. Figure 1.1 shows the relationship between *the Guidebook* and the seismic investigation.

What needs to be special pointed out is that the seismic investigation has significant scientific properties, and there are always some uncertainties in the practical

Table 1.2 Seismogenic mechanism investigation

Sequence number	Investigation objectives	Investigation approach	Contents	Expected outcomes
1	Determine the deep geophysical environment of earthquake, providing a basis for the interpretation of earthquake genesis and tectonics	Research on velocity structure of focal area	By processing and analyzing of seismic observation data, obtain a complete and accurate four-dimensional active image, and carry out the research of seismogenic environment in the deep source zone	Detailed three-dimensional seismic velocity structure of focal area
2		Research on electrical structure of focal area	Through processing and analyzing of magnetotelluric data, carry out lithospheric magnetotelluric inversion in focal area	Three-dimensional electric structure of lithosphere in focal area
3		Research on magnetic field structure in focal area	Through geomagnetic data processing and analysis, carry out the inversion for lithospheric magnetic field structure in the focal area	Spatial three-dimensional magnetic structure of lithosphere in focal area
4		Research on density structure of focal area	Through the processing and analysis of gravity observation data, interpret the gravity variation images and anomalies, and obtain lithospheric density and its dynamic evolution in the focal area	Two or three dimensional density fine structure of crust in focal area
5		Artificial seismic detection	Conduct wide angle refection/refraction, high-resolution refraction profiles and deep seismic sounding, and analyze seismic velocity structure, crustal stratification structure and fault morphology in the deep and shallow parts of the crust	Fine velocity structure of the crust in the focal area, deep and shallow tectonic relationship, and location, type and spatial geometry distribution of faults

(continued)

Table 1.2 (continued)

Sequence number	Investigation objectives	Investigation approach	Contents	Expected outcomes
6		Comprehensive interpretation of the mechanisms of earthquake generation	Based on research results of the above four aspects, obtain the general conclusions and scientific understanding about thc causes of earthquake, and analyze shortcomings of current schemes of earthquake monitoring and scientific investigation	Comprehensive explanation of seismic generation, and suggestions for further research and strengthening seismic work

work. Therefore, *the Guidebook* cannot cover all aspects. In the process of scientific practice, it is quite needed to analyze the specific issues according to the actual situation of the earthquake zone, adopt flexible and site-specific principle, combine the scientific feature and operability and give attention to the timeliness of scientific investigation and respond to government and society needs actively. In addition, more attention should be paid to the quality of field data in specific work, and the specific technical requirements proposed in each work in *the Guidebook* should reach high standards (parameters) and not lower than the minimum standards (parameters) as far as possible.

It can be believed that, with the persistent development of earthquake science and technology and scientific investigation practice in the future, *the Guidebook* will continue to improve.

Table 1.3 Earthquake evolution process investigation

Sequence number	Investigation objectives	Investigation approach	Contents	Expected outcomes
1	Identify the physical and chemical anomalies in the process of earthquake generation and provide a basis for the scientific interpretation of seismic observational anomalies	Retrospective study on "meta-instability"	Through the analysis of seismic observation data by multiple approaches, carry out the "meta-instability" retrospective test, obtain anomaly information before earthquake and explain observed anomaly and its evolution process	Physical interpretation of seismic observation anomalies and their evolution
2		Research on characteristics of fault locking before earthquake	Through the processing and analysis of GNSS, InSAR, leveling and other deformation data, obtain the characteristics of deformation field and fault motion in the epicenter and surrounding areas before earthquake, and carry out inversion research of fault locking and slip deficit	Fault locking area and its locking degree before earthquake
3		Research on anomaly geophysical observation	Through the analysis of geophysical observation data in the epicenter and surrounding areas, obtain the characteristics of preseismic anomalies and their co-seismic and post-seismic variations	Spatio-temporal evolution of geophysical anomalies in the epicenter zone and surrounding areas
4		Research on geochemical anomaly observation	By sorting and analyzing the geochemical observation data in the epicenter and its surrounding areas, obtain the characteristics of geochemical temporal-spatial anomalies	Influence of deep fluid migration and release intensity in the epicenter its and surrounding area
5		Comprehensive interpretation of earthquake generation process	By combining research results of the above four aspects and numerical simulation analysis, explain seismic anomaly change process, and obtain overall conclusion and scientific understanding about earthquake generation process, and analyze the shortcomings of current seismic monitoring and scientific investigation	Comprehensive interpretation of earthquake generation process and suggestions for further research and strengthening seismic work

Table 1.4 Influence of the studied earthquake on regional earthquake hazard analysis

Sequence number	Investigation objectives	Investigation methods	Contents	Expected outcomes
1	Analyze the influence of investigated earthquake on regional earthquake risk, and study earthquake potential in the epicenter and its surrounding areas	Determination of seismic sequence evolution features	Based on seismic catalog, calculate the basic parameters of seismic sequence, determine sequence types and development trends and analyze fault structures	Statistical seismological models of seismic sequence
2		Research on variations of regional gravity field	Through the analysis of gravity data in earthquake area, obtain the image of gravity change after earthquakes and invert for the density of medium, gravity disturbance and dynamic evolution in focal area	Characteristics of time-varying regional gravity field
3		Research on variation of regional deformation field	Through the processing and analysis of GNSS, InSAR, leveling and other deformation data, obtain characteristics of post-seismic regional deformation field, and analyze effect of seismic fault rupture on the deformation loading or unloading on surrounding faults	Evolution characteristics of regional post-seismic deformation field, give the influence characteristics of the seismic fault on the regional fault activity
4		Research on regional stress evolution	Through seismic data processing and analysis, calculate regional stress field and fault Coulomb stress changes	Characteristics of regional stress field and fault Coulomb stress change
5		Comprehensive interpretation of regional earthquake risk variation	By synthesizing the research results of the above four aspects, obtain the general conclusion and scientific understanding about the regional earthquake risk change, and analyze the insufficiency of earthquake monitoring and scientific investigation	Comprehensive interpretation of regional seismic potential changes and suggestions for further research and strengthening earthquake preparedness

Table 1.5 Earthquake disaster mechanism investigation

Sequence number	Investigation objectives	Investigation methods	Contents	Expected outcomes
1	Determine the mechanism of strong ground motion and earthquake disaster and put forward measures to reduce earthquake disaster losses	Research on strong ground motion of engineering sites	Through the processing and analysis of ground motion observation data, obtain the distribution law of strong ground motion, and conduct site effect study and numerical simulation of ground motion	Ground motion influence field and earthquake effect mechanism
2		Research on seismic response of engineering structures	Obtain the seismic damage characteristics of engineering structures by processing and analyzing strong motion data of engineering structures	Mechanism of engineering structure disaster
3		Research on seismic response of major and special projects	Through investigation of earthquake disasters in major and special projects, analyze the process of damage and disaster	Mechanism of major and special project disaster
4		Research on seismic geological hazards and secondary hazards	Through remote sensing and field investigation, obtain the spatial distribution of seismic geological disasters and secondary disasters, and explain the causes of disasters	Occurrence mechanism of seismic geological disasters and secondary disasters
5		Comprehensive interpretation of earthquake disaster mechanism	By combing the research results of the above four aspects, obtain overall conclusion and scientific understanding of earthquake disaster mechanism, and analysis the weakness in current related work	The process and its mechanism of earthquake disaster, and the following researches and seismic mitigation measures

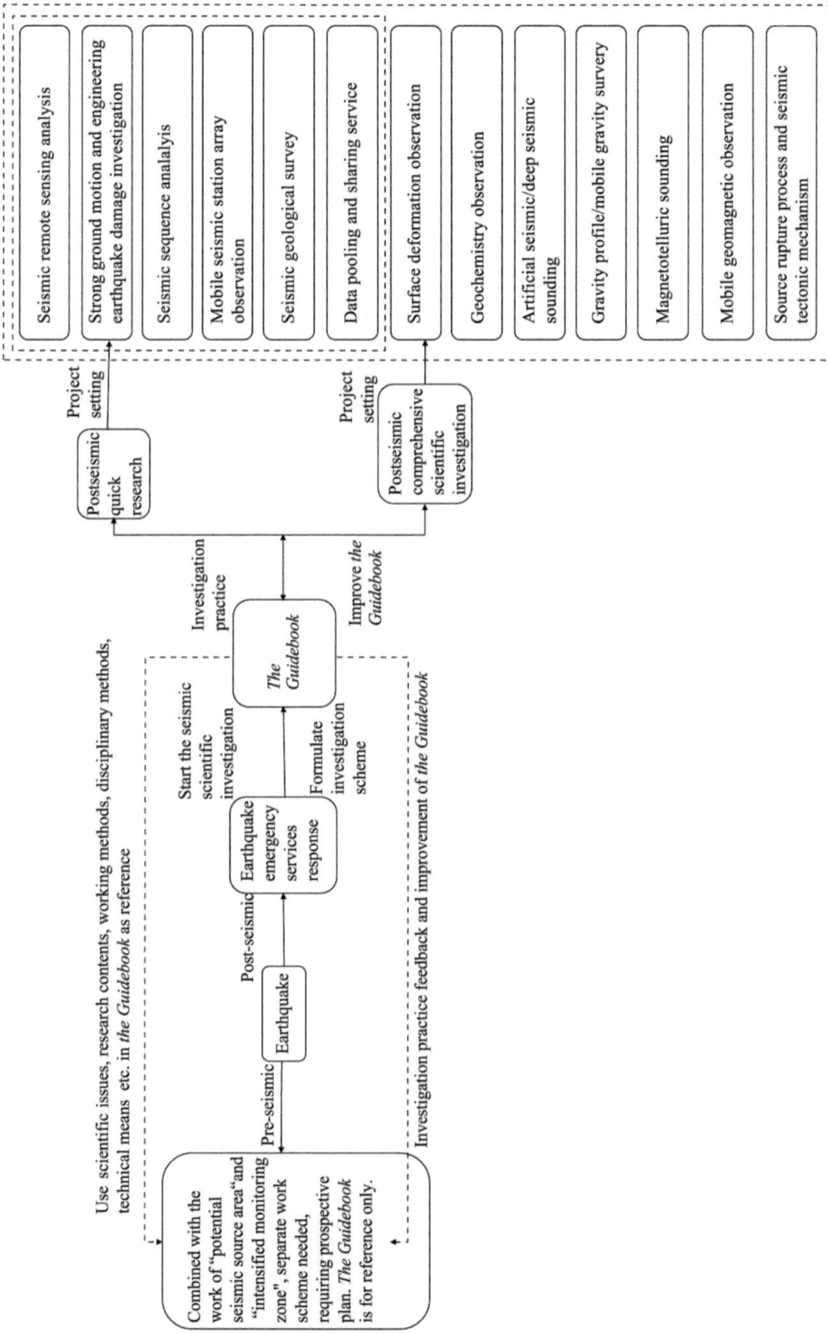

Fig. 1.1 Schematic diagram of instructions of *the Guidebook*

Chapter 2
Procedures for Earthquake Scientific Investigation

According to the requirements of *Earthquake Emergency Service Response Level (Interim) of China Earthquake Administration*, when an earthquake of $M \geq 7.0$ occurs in densely populated areas and an earthquake of $M \geq 6.0$ occurs in Beijing-Tianjin-Hebei area, large cities in the Yangtze River Basin, Pearl River Delta and provincial capitals, Level 1 emergency service response will be activated. When an earthquake of $6.0 \leq M < 7.0$ occurs in densely populated areas or an earthquake of $5.0 \leq M < 6.0$ occurs in Beijing-Tianjin-Hebei, large cities in the Yangtze River basin, Pearl River Delta and provincial capitals, level II emergency services shall be activated. According to the requirements of the "List of primary and secondary response actions for post-earthquake emergency services of the China Earthquake Administration within 12 h after the earthquake", the Department of Science and Technology of the China Earthquake Administration starts the scientific investigation within 8–12 h after the earthquake and organizes the preparation of the plan of earthquake scientific investigation (referred to as "earthquake investigation").

2.1 Indoor Preparation Stage

1. Data preparation

(1) Collect all kinds of rapid output results of the earthquake, such as seismic parameters, seismic sequence, focal mechanism solution, focal rupture process, post-earthquake deformation.
(2) Collect seismic and geological data (fault structure, geological maps of different scales, topographic maps, etc.), geophysical fields (artificial earthquakes, gravity profiles, campaign gravity, magnetotelluric, etc.), deformation, remote sensing images (high-resolution satellite images and aerial images, etc.), geochemistry,

Institute of Earthquake Forecasting, China Earthquake Administration,
A Guidebook to Earthquake Scientific Investigation,
https://doi.org/10.1007/978-981-96-0418-0_2

hydrogeology, geomagnetic data and other data related to the area where the earthquake occurred.

(3) Formulating the content, focus and route of field work, etc.
(4) It is suggested that relevant organizations should collect and sort out the basic data and materials of earthquake science in the earthquake risk area and even the whole country on a daily basis, so as to prepare for the rapid acquisition of relevant data and materials in the post-earthquake scientific investigation work.

2. Equipment and labor protection equipment

(1) Various field observation instruments and tools, mainly including handheld GPS, geological compass, laser rangefinder, geological hammer, ice axe, differential GPS, 3D laser scanner, UAV airborne LiDAR and GNSS observation instrument, geophysical field exploration related instruments, etc.
(2) Preparation of all kinds of field work equipment and labor protection supplies.

2.2 Stage of Field Investigation and Observation

According to the general requirements of the scientific research program, *the Guidebook* is applied to carry out the corresponding field investigation and observation.

(1) In the field work stage, the relevant laws and regulations of national production safety and industry standards should be strictly observed. In accordance with *the Work Safety Law of the People's Republic of China, the Road Traffic Safety Law of the People's Republic of China, the Labor Law of the People's Republic of China, the National Security Law of the People's Republic of China, the Law of the People's Republic of China on Guarding State Secrets* and other national laws and regulations, as well as relevant provisions of the Ministry of Emergency Management and the China Earthquake Administration, and in light of the actual conditions of the scientific investigation, make safety plans.
(2) Due to the difficult working conditions, urgent time and heavy tasks, field work should be divided into several groups, which should strengthen communication and carry out related work in accordance with unified working methods and technical approaches.
(3) Field work should be gradually spread out and form a system in the way of gradual execution of observational points, lines and surfaces.
(4) The work of the survey sites and observation sites should be steadily carried out, and the data should be as rich as possible and recorded accurately.
(5) On the basis of actual investigation, typical sites and phenomenon points should be selected as the key working points of relevant means.
(6) The field work data should be carefully organized and comprehensively analyzed, and the daily summary should be made so that the field work summary report can be prepared in time after the end of the field work.

2.3 Indoor Analysis and Summary Stage

(1) Indoor analysis and identification of dating samples, rock samples and geochemical samples.
(2) Analysis and processing of various observational and sounding data (geophysical, geochemical, deformation, etc.).
(3) Drawing and mapping of remote sensing interpretation map, geological and geomorphic plan, geological and geomorphic profile map, surface rupture distribution map, seismic structure map, fault profile map, trench profile map, geological disaster distribution map, crustal deformation distribution map, observational anomaly distribution map, engineering earthquake damage distribution map and observation results map by means of various disciplines.
(4) Preparation of special reports.

2.4 Results and Data Archiving

1. Contents and forms of results

Field observational data, sounding data, test data, drawings, photos and images of each subject.

2. Main contents of the summary report

The report shall include but not be limited to the following contents:

(1) Preface.

The preface includes the work objectives, work content and workload, survey methods and work deployment, list of working groups and division of labor, etc.

(2) Data acquiring, analysis and processing.

Including special investigation, measurement data analysis, processing, etc.

(3) Summary of achievements.

The scientific research results should be summarized comprehensively, the work results are compared with the expected objectives, and the existing problems and deficiencies are put forward to provide reference for the future scientific research.

3. Original data submission and archival

Raw data include, but are not limited to, the following:

(1) Basic geological, mapping and geographic information related to each topic group.

(2) Various original measurement (surveying and mapping) data obtained from the field work of each special research group.
(3) Test and analyze data of each special subject group.
(4) All kinds of field work photos (sets), video data and text descriptions, etc.
(5) Confidential information shall be filed and managed in accordance with the relevant national security management measures.

Chapter 3
Seismogeology Investigation

3.1 Objectives

The study of the geometric and kinematic characteristics in the earthquake surface rupture zones, co-seismic surface displacement and distribution of seismic hazards based on the high-resolution remote sensing image analysis is the most important goal for the seismic and geological survey. It will provide an important reference the scientific decision-making in the determination of seismic structures, earthquake disaster loss evaluation, earthquake trend prediction and post-disaster reconstruction.

3.2 Contents

Firstly, the seismic structures and structural settings in the earthquake zones can be carried out through a survey of earthquake surface rupture. Secondly, in order to obtain surface deformation and large-scale mapping, a seismic and geological survey and measurement along the seismic faults based on the high-resolution remote sensing image analysis is necessary. Lastly, to obtain quantitative parameters of fault activity, the geological and geomorphologic field surveys, identification of fault activity and study on the long-term activity behavior of faults should be carried out in the earthquake areas.

1. Field survey of earthquake surface rupture zone

Earthquake surface rupture zone is a general term for surface rupture and deformation of seismic faults in large earthquakes. In general, an earthquake surface rupture zone shows a certain width. It often consists of seismic faults, earthquake bulges, earthquake fractures and earthquake trenches.

1) Survey area of earthquake surface rupture.

Institute of Earthquake Forecasting, China Earthquake Administration,
A Guidebook to Earthquake Scientific Investigation,
https://doi.org/10.1007/978-981-96-0418-0_3

Earthquake surface rupture can be preliminarily identified through the high-resolution remote sensing image analysis in post-earthquake. For the exposed seismic faults, surface rupture can be tracked along the fault strike and dip. For the surface rupture caused by a hidden active fault, the area of field survey should be extended according to the actual situation.

2) Survey of seismic fault properties.

Based on the observation of the fault profile and the analysis of surface rupture, the seismic fault properties including positive fault, reverse fault and strike-slip fault can be identified, and attitude of fault should be indicated on the related maps.

3) Survey of structural and deformational type, geomorphologic shape and geometric distribution of earthquake surface rupture zone.

Firstly, the structural type, deformational type and geomorphologic shape in the different sections along the seismic fault need to investigate in detail. Secondly, the fault property can be combined in the analysis of kinematic characteristics of fault. To reflect the plane distribution, rupture segmentation and structural type of the earthquake surface rupture zone, large-scale geological mapping should be carried out according to the actual situation.

4) Survey and displacement measurement of earthquake surface rupture zone.

Firstly, the plane fault marks reflecting the horizontal displacement and the section fault marks reflecting the vertical displacement should be found in detail. And then the displacement of seismic fault in different sections is investigated comprehensively and the horizontal and vertical displacement is measured accurately. Lastly, displacement maps of seismic faults along the fault strike are mapped.

5) Survey and analysis of earthquake-associated structures.

A series of associated structures of extension and compression are surveyed, including small pull-apart basins, half grabens, faulted upfolds, bulges and ridges. Based on the morphological characteristics of associated structures, the kinematic characteristics of seismic structures and regional stress environment are analyzed comprehensively.

6) Survey of non-structural fractures.

Geometric shape, distribution and displacement of non-structural fractures including earthquake-induced surface fractures (i.e., landslides, potholes) are surveyed and distinguished them from the earthquake surface rupture during the survey.

7) Characteristics and comprehensive analysis of surface deformation zone caused by the earthquake.

Firstly, based on the field survey, the geometric distribution, macroscopic scale, deformation characteristics and displacement distribution of the earthquake surface rupture zone can be surveyed. Secondly, the surface rupture length and maximum co-seismic displacement are checked, and the geometric distribution map and co-seismic displacement distribution map of the earthquake surface rupture zone are mapped with relevant captions. Thirdly, based on the survey of earthquake surface rupture and mapping and the collection of surface deformation, the deformational and damaged characteristics, and the influence of geological and geomorphic conditions on deformation and damage should be analyzed. Lastly, the relationship between characteristics of surface deformation and seismic structures is studied.

2. Survey of the seismic environment

Survey of the seismic environment refers to the survey and quantitative study of active structures in the earthquake area based on the collection of background, information including regional geology and deep structure. The main contents of the survey of seismic environment include the quantitative study on the late Quaternary activity of seismic faults and the late Quaternary activity investigation of other related faults in the earthquake area.

1) Study on long-term active behavior of seismic faults.

Firstly, the active history of faults in different periods is analyzed, and the long-term active behavior of seismic faults is studied to identify the recurrence of large earthquakes based on the detailed field survey. Secondly, measurement of faulted landform and paleoseismic trenches can be carried out. Thirdly, the Quaternary dating in the key strata and landform surfaces is carried out by taking necessary geological dating samples. Lastly, the quantitative parameters of the late Quaternary activity are obtained, including the kinematic properties, attitudes, scale, segmentations and slip rates of seismic faults and recurrence models and the cycle of paleoearthquakes.

2) Survey of other active structures in the earthquake area.

Based on the collection of information, the Quaternary activity of active structures in the earthquake area can be preliminarily determined. For the more studied active faults, the quantitative parameters should be obtained as much as possible, including fault property, attitude, segmentation, active age, slip rate, recurrence and intervals of paleoearthquake. For the faults with less studied and unknown late Quaternary activity, the necessary supplementary survey and identification of fault activity should be carried out. If there are other active structural types in the earthquake area in addition to active faults, relevant surveys should be conducted. Combination characteristics of active structures in the earthquake area are studied, and then the property, type, distribution, active age and late Quaternary activity are also obtained. In volcanic areas, the characteristics of volcanic activity should be investigated, including the property, scale, activity age and the latest volcanic time. Furthermore, the relationship between historical earthquakes and volcanic activity in the area is investigated, and the possibility of volcanic activity caused by post-earthquakes also should be analyzed.

3) Comprehensive analysis and determination of seismic structure.

Detailed field seismic and geological survey should be carried out in the exposed area of seismic faults, and necessary geophysical exploration, drilling and trench shall be carried out in the hidden fault area to find seismic faults. Furthermore, the geometric distribution, property, scale, active age, paleoseismic characteristic and active segmentation of seismic fault are studied deeply. Based on the above work, combined with the seismic deformation, geochemical and seismic data including deep structure, focal mechanism, fine location of small earthquake and near-field seismic measurement, seismic structure are comprehensively determined. Lastly, on the basis of the condition of sufficient data, the seismic structure model is established, and then the seismic model, recurrence type and recurrence interval of large earthquake on the seismic structure is studied.

4) Determination of seismic gaps and seismic risk in the seismic structure zone (or block boundary zone).

Large earthquake usually occurs in large active fault zone (or block boundary zone), and a single earthquake often appears as the earthquake rupture in some sections of fault. Based on the basic theory of segmentation and characteristics of segmental fracture in the active fault, the survey of earthquake surface rupture, the long-term activity behavior of seismic fault and the comprehensive analysis and determination of seismic structure, the section of the seismic structure zone (or block boundary zone) that has not been ruptured for a long period in the history is determined, namely the determination of seismic gaps. Lastly, the seismic risk is determined by analyzing its potential trend of seismic fault in the future.

3. Survey of seismic and geological hazards

Seismic and geological hazards refer to the hazard caused by the deformation or damage of the geological body under the action of the earthquake. Survey of seismic and geological hazards are mainly to study earthquake falling, landslides, dammed lakes, debris flows, ground fissures, liquefaction and earthquake-induced subsidence of soft soil based on the comprehensive collection of data related to seismic and geological hazards including topographic maps, geological maps, geomorphological maps, remote sensing images, historical seismic and geological hazards, stratigraphic distribution, engineering geology, hydrogeology, geodetic survey and strong ground motion. In addition, geometric shape, distribution and displacement characteristics of non-structural fractures including earthquake-induced landslides and subsidence are surveyed and distinguished them from the surface rupture of structural earthquakes. Finally, the shape, scale and spatial distribution of various types of seismic and geological hazard are determined, and then the distribution maps of seismic and geological hazard and local feature maps of key hazard points are mapped.

3.3 Methods and Basic Requirements

The main technical methods and basic requirements of seismic and geological survey are as follows:

1. High-resolution remote sensing image analysis

1) Data preparation and basic requirements.

(1) The high-resolution remote sensing image data in pre- and post-earthquake covering the earthquake area should be obtained immediately after the earthquake. The spatial resolution of remote sensing image data should be better than 5 m, and the resolution in local key area should reach the submeter level (< 1 m).
(2) As far as possible, multi-type and multi-temporal image data should be acquired and screened. Try to choose clear images with few (< 5%) clouds and cloud shadows. Spring and autumn data should be selected for preearthquake images in northern China, winter data should be selected for low-latitude areas with dense vegetation cover and summer data should be selected for alpine areas with snow cover.
(3) Early historical aerial or satellite images should be collected in densely populated cities and surrounding areas where the transformation of human activities is great. Radar remote sensing images should be collected in plain areas and basin areas.
(4) The DEM topographic data in the earthquake area can be obtained. The spatial resolution of DEM data used for orthographic correction of high-resolution images should be better than 25 m or no less than 1:50,000 scale.

2) Data processing.

(1) Image preprocessing.

The remote sensing image is processed by geometric precision correction, registration, uniform color, mosaic and other processing, and the data are projected to a unified coordinate system. In the mosaic processing of multi-scene images, if the grayscale between two adjacent scene images in the same channel show a greater contrast, it is appropriate to carry out uniform color processing. Inlaid parts should not appear as obvious blur or double shadow and should not appear as special acquisition object dislocation.

(2) Enhancement processing of image information.

According to the spectral channel and interpretation requirements of the remote sensing data, the appropriate band combination is selected to carry out the next work. The gray image of panchromatic band with higher resolution should be fused with the multi-spectral color image with lower resolution to obtain the color image with the same resolution as panchromatic band. Different types of surface features are highlighted by the combination of true-color images and false-color images in different wavebands.

By means of difference calculation, ratio calculation and band stretching between bands, the difference of ground feature spectra between different bands is highlighted, the contrast is increased, the vegetation characteristics, soil water-rich difference and microgeomorphic changes is highlighted and linear features associated with active fault information is enhanced.

For some seismic structures with weak information, specialized image enhancement and spatial enhancement methods can be used to enhance image interpretation.

3) Data analysis and interpretation.

In the epicenter area, high-resolution remote sensing analysis in the geology, geomorphology and major structures is carried out using the preearthquake data. Based on the data obtained post-earthquake, the earthquake surface rupture zone, seismic structure and seismic and geological hazard can be preliminary interpreted. And then the macro-scale of the earthquake rupture and the spatial characteristics of the hazard distribution are preliminary determined and analyzed, which provides a reference for the detailed field geological survey.

(1) Elements of remote sensing interpretation.

The elements of remote sensing interpretation mainly include geological elements, geomorphic elements and active structural elements. Geological elements mainly include stratigraphic element. Stratigraphic units should be interpreted to distinguish loose deposits, sedimentary rocks and magmatic rocks. Geomorphic elements mainly include regional planation surface, collapse, landslide, landslide dam, downland, water system, gully, river terrace, alluvial fans and proluvial fans. Active structural elements mainly include surface rupture zone, pull-apart basin, fault scarp, triangular facets, reverse scarp, trough, sag pond, faulted terrace, faulted proluvial fan, beheaded channel, betrunked channel, pressure ridges, bulge, fold and en-echelon surface fracture.

(2) Methods of interpretation.

Intuitive and obvious elements can be used to directly determine by the visual interpretation. These elements mainly include location, morphological structures and activity of structural geomorphic elements, such as linear scarps, pull-apart basins, river terraces, alluvial fans and proluvial fans, beheaded channels and betrunked channels.

For the regions or elements whose information is not very intuitive, according to the relationship between remote sensing image features, the method of tracing from macro-information to local information, extending from peripheral features to the central target, and inferring from clear features to fuzzy information is adopted. And then according to the tone anomalies, color anomalies, anomalies of texture and structure features in the remote sensing image, the approximate location of the active fault and other information can be determined.

2. Field geological and geomorphic survey and mapping

1) Working methods.

The two traditional methods used in field surveys are the tracing method and the crossing method. In general, it is advisable to use the tracing method as the main method and the tracing method as the auxiliary method for the survey of Quaternary geomorphology. For linear structures, the tracing method is the most commonly used method and the crossing method is the auxiliary method, such as earthquake surface rupture zone, seismic fault and active fault. Generally, fold structures with large widths should be combined with the tracking method and crossing method. The tracing method should be used in the surveys in the Quaternary stratigraphy and volcanos.

2) Basic requirements.

(1) Based on the distribution of surface rupture and other phenomena, the seismic structure and the survey of seismic and geological structure is carried out.
(2) Banded geological mapping and measurement of co-seismic displacement is carried out for the seismic surface rupture zone at the scale of 1:50,000–1:1000. The mapping work involved in field survey mainly includes the mapping of the earthquake surface rupture zone, and in some cases, it may also involve the mapping of the active faults in the earthquake area. The earthquake surface rupture zone generally has a certain width. Referring to the relevant technical specifications of active fault mapping, it is suggested that the mapping area of the rupture zone should follow these principles.

If the earthquake surface rupture zone involves a single active normal fault, the banded mapping area should extend 1–2 km on both sides of the seismic fault. If there are two or more parallel seismic faults, the banded mapping area should extend outward 1–2 km on both sides of the fault trace. If the rupture zone involves reverse fault and no active fold, the mapping area is the same as that of the area of normal fault. If there is an active fold, the mapping area should cover the active fold zone and extend outward 1–2 km. If the rupture zone involves a single strike-slip fault, the mapping area should be 1–2 km outward extension of the earthquake fault trace. If there are two or more parallel seismic faults, the mapping area should be 1–2 km outward extension of the earthquake fault trace. If there are pull-apart basins or compressional uplifts in the fault relay, the mapping area should cover all the pull-apart basins or compressional uplifts, and extend outward 1–2 km.

(3) The general deformation characteristics including spatial distribution and geometric structure of the earthquake surface rupture zone and the structural phenomena reflecting the kinematic properties of the seismic faults are photographed in detail, and the necessary text and audio-visual descriptions should be made. The structural phenomena mainly include geometric combination of secondary faults, marked geological body, displacement of geomorphic surfaces or structural lines and typically associated structures.

(4) For the marked geological body, geomorphic surface or structural lines, necessary age samples or specimens should be collected during the earthquake. Furthermore, the longitude and latitude of marked geological bodies, geomorphic surfaces or structural lines can be measured with portable GNSS receiving equipment and plotted on the banded geological map of the earthquake surface rupture zone.
(5) Before the completion of the field survey, a detailed review of the original data should be conducted and include the following.

① Field records check, including the integrity and accuracy of observation sites, descriptions and charts of observed phenomena, sample ID and sampling location.
② Samples collection. The aim is to determine whether additional samples are sampled.
③ Verification of important plans and sections, including determining whether the banded geological map of the earthquake surface rupture zone and measured sections are complete and accurate.

3) Elements of survey.

(1) The location, strike and width of the earthquake surface rupture zone are surveyed, and its distribution is further determined.
(2) The types, deformed styles and combination characteristics of surface fractures are surveyed. Furthermore, associated structures mainly include graben, fold uplift, extrusion depression, bulge and pressure ridge. Finally, the kinematic properties of seismic faults are determined.
(3) Reliable displacement markers within the rupture zone and co-seismic displacement are measured.
(4) The location, attitude, property and displacement of the fault are surveyed.
(5) The Quaternary geomorphic types, distribution and scale are surveyed.
(6) The structural geomorphology and micro-geomorphology, type, distribution and deformation characteristics along the fault is surveyed, mainly including triangular facets, fault scarp, trough, sag pond, terraces, rivers, alluvial fans, proluvial fans and beheaded channels.
(7) To determine the parameters of faulted geomorphology, the geomorphic surface along the fault should be surveyed and further graded.
(8) The distribution of the Quaternary strata is surveyed, and stratification, lithology, color, thickness, attitude and genetic type of strata are determined. Furthermore, necessary dating samples for absolute dating should be collected.
(9) Relationship between fault and stratum or geomorphic surface cutting and compaction is surveyed.

① The structure, material composition and width of the fault zone are surveyed.
② To study seismic faults and paleoseismic events, shallow geophysical exploration and trough excavation can be carried out in the area of suspected seismic faults

and hidden faults. If the trough cannot be excavated, the drilling verification can be carried out possibly to determine the buried depth of the up-breakpoint and the fault dislocation.

③ The type and attitude of fold structure and the stratum involved in deformation by fold are surveyed.

④ The distribution of the Quaternary volcanic rocks is surveyed, and the types of rocks, color, thickness, intrusions or ejection periods and internal structure are further determined.

3. Measurement of co-seismic displacement and structural micro-geomorphology or geomorphology

1) Measurement methods.

In the field, co-seismic displacement can be measured directly by choosing significant and reliable faulted landform markers using a compass and tape. Similar to the measurement of structural micro-geomorphology or geomorphology, the co-seismic displacement can be obtained indirectly by using high-precision measurement means for profile and topographic measurement.

2) Measurement of co-seismic displacement.

The geomorphic and geological body, plant, mark of animal and human activity and artificial facilities which is faulted in the earthquake surface rupture zone can be used as measurement marker. The measurement markers mainly include river valley, channel, mountain ridge, alluvial and proluvial fan, strata, rock body, tree roots, roads and structures. Furthermore, the type of displacement marker should be recorded during measurement, and different measurement means should divide into classes of corresponding precision.

3) Technical requirements for measurement of faulted landform.

(1) Section measurement should meet the following requirements.

① The starting position and direction of the faulted section are determined.

② The plane coordinates and elevation values of ground points or other measurements that can be converted to three-dimensional spatial coordinates are measured.

③ The measurement route can be composed of several line segments which are connected end to end. The starting points, end points and inflection points of the line segments should be selected in the flat terrain, and markers should be placed as needs.

(2) The topographic survey should meet the following requirements.

① The measuring area is determined.

② The plane coordinates and elevation values of discrete points on the ground, other data that can be converted to discrete points on the ground is measured, such as aerial photographic data.

(3) Measurement of vertical faulted landform.

① Geomorphology formed by vertical fault activity should be selected as the measurement objects, including fault scarps and folds.
② Vertical faulted landforms can be measured by section measurement or landform measurement.
③ When measuring the fault scarp, the landform surface on both sides of the fault scarp should be the same landform surface, and the section direction should be perpendicular to the strike of the fault scarp possibly.
④ The section crosses the original landform surface above and below the fault scarp, and the length of the section line on both sides of the scarp can determine the original landform surface at the top and bottom of the scarp.
⑤ The trend lines of the original landform surface of the top and bottom of the scarp are determined by linear fitting or manual interpretation, and the vertical distance between the two trend lines is the vertical displacement.
⑥ When using the profile measurement method to measure folds, we should ensure that the profile direction is perpendicular to the fold axis and the profile crosses the fold belt and the original landform surface on both sides, the length of the section can be used to determine the landform surface on both sides of the fold.
⑦ The trend line of the landform surface outside the fold area is determined by linear fitting or manual interpretation, and the vertical distance between the trend line and the top of the fold is the vertical lift.
⑧ When the landform survey method is used to measure the vertical faulted landform, the area of landform measurement should cover the selected scarp and fold and the profile data of the fault scarp and fold should be extracted from the landform measurement data.

(4) Measurement of horizontal faulted landform.

① The landform controlled by horizontal fault activity should be selected as the measurement objects, such as faulted channels, diluvial fans and terraces.
② The topographic measurement should be used for horizontal faulted landform survey.
③ The trend lines of the same linear marks on both sides of the deformation area are determined by linear fitting or manual interpretation methods. The linear marks mainly include the middle line of the bottom of the channel, the edge of the terrace at the same level as the river, the middle line of the river and other linear features. The distance between the trend lines along the fault strike is measured, and the distance is horizontal displacement of the fault.

4) Commonly used high-precision measuring instruments.

Common high-precision measuring instruments include total station, differential GNSS, UAV, 3D laser scanner and airborne LiDAR. It can be used reasonably according to the actual situation of the earthquake site and equipment.

4. Shallow geophysical exploration

1) Common means and scope of application.

Shallow geophysical exploration generally includes shallow artificial seismic detection, high-density electrical detection, geological radar detection and other means.

Shallow artificial seismic exploration and high-density electrical detection are generally applicable to hidden faults buried several meters to several hundred meters deep at the upper breakpoint. When detecting hidden faults with buried depths of several meters to tens of meters, the S-wave (SH wave) reflection prospecting method should be adopted. When detecting hidden faults with buried depths of tens to hundreds of meters, the P-wave reflection prospecting method should be adopted. When the seismic and geological conditions are complicated, the refraction and reflection wave combined exploration method should be adopted. Furthermore, geological radar detection is generally applicable to the buried depth of the upper breakpoint of several meters to tens of meters of hidden fault. Generally, shallow artificial seismic exploration has the widest applicability and better detection effect; it is recommended as the preferred exploration method.

2) General requirements for shallow artificial seismic exploration.

(1) Before shallow artificial seismic exploration, field tests and parameter demonstration should be carried out to determine the best seismic wave excitation, reception mode and conditions and the best parameters for the observation system.
(2) The detection profile should be as perpendicular as possible to the strike of the target fault. The location of the measuring lines should be chosen in the area with less topographic relief. When the topography along the measuring lines changes greatly, the position coordinates and elevation of the shot points and geophones should be measured, and the sketch of the topographic profile along the measuring lines should be mapped.
(3) Seismic lines should be laid in a straight line. If a straight line cannot be laid in a straight line due to site conditions, the line can be turned, but the turning angle should not be too large. Measuring lines should coincide as far as possible in the joint exploration of compressional and shear waves or joint exploration of refraction and reflection waves.
(4) The end points, inflection points, shot points and geophones of the measuring lines should be located by instruments, and the flatness error of position is less than 3 m and the elevation error of position is less than 1 m. When exploration in water, GNSS should be used for the position of the seismic lines and comply with the requirements of GB/T 18314-2009.

(5) A multiple coverage observation system is used for shallow wave reflection exploration and the channel spacing should satisfy the spatial sampling theorem. When the fault is accurately located, the two-dimensional or three-dimensional observation method with no more than 2 m channel spacing should be adopted, and the coverage should be determined according to the field test results.
(6) The correlative meeting and pursuing observation system is applied to the shallow refraction wave exploration. There are no less than 4 geophones in the meeting segment to receive the refraction waves from the tracked interface, and the minimum offset is 0 m and the group interval can be 5–10 m.

5. Palaeoseismic trench

1) General work flow.

Paleoseismic trench investigation includes seven stages: preliminary preparation, site selection, topographic mapping and trench excavation, trench records, trench interpretation and paleoseismic identification, sampling and dating and results. Furthermore, trench interpretation, sampling and dating and paleoseismic identification can be carried out simultaneously.

2) Main content.

(1) Preliminary preparation mainly includes basic data collection, image interpretation, field survey and data storage.
(2) Location of the trench. Location of the trench should select suitable sites according to the normal faults, reverse faults and strike-slip faults for the excavation of the trench.

When excavating a trench on normal faults, it is recommended to select the site with a single scarp, continuous accumulation in front of the scarp and the cumulative height of which is double or more than the co-seismic displacement. The trench can be excavated on the fault scarp on alluvial-proluvial fans, the second or third-grade river terraces and the parts with small graben structures. When excavating a trench on a reverse fault, it is recommended to select the site with a single scarp, continuous accumulation in front of the scarp, and the cumulative height is double or more than the co-seismic displacement. The trench can be excavated on the fault scarp on alluvial-proluvial fans, the second or third-grade river terraces, and the reverse scarp or depression which is caused by thrusting. When excavating a trench on a strike-slip fault, it is recommended to select the pull-apart basins and sag ponds and the channel (less than 5 m in width) and other micro-landform which is faulted.

(3) Topographic mapping and trench excavation. Firstly, the original landscape at the excavation site should be recorded by taking pictures. Secondly, the range of topographic mapping should cover the trench and the nearby structural deformation landform. Thirdly, it is recommended that faulted landform measurements should be carried out in accordance to the specification of "*Active fault survey—Measurement of faulted landform*" (DB/T 71-2018) near the trench before the

trench excavation. Fourthly, during the excavation of the trench, when the depth of the trench is greater than 2 m, supporting measures should be taken to prevent the collapse of the trench. Lastly, when the excavation depth is greater than 3.5 m, it is appropriate to adopt the step excavation method.

(4) Other related works, such as trench record preparation, trench interpretation, paleoseismic identification, sampling, dating and results.

6. Technical requirements

More technical requirements including data processing, interpretation and thematic map production in the high-resolution remote sensing interpretation are suggested for referring to "*Mapping of active fault in 1:50,000*" (DB/T 53-2013), "*Remote sensing technique in regional geology survey*" (1:50,000) (DZ/T 0151-1995), "*Active fault survey—Remote sensing investigation*" (DBT 69-2017) and "*Active fault survey—Field geological investigation*" (DB/T 82-2020).

More technical requirements including field observation requirements, data sorting, results and mapping in the field geological and geomorphic survey are suggested for referring to "*Post-earthquake field works—Part 3: Code for field survey*" (GB/T 18208.3-2011), "*Surveying and prospecting of active fault*" (GB/T 36072-2018) and "*Active fault survey—Field geological investigation*" (DB/T 82-2020).

More technical requirements including measurement data requirements, data inspection, data processing and mapping of relevant instruments in the measurement of faulted landform re suggested for referring to "*Active fault survey—Measurement of faulted landform*" (DB/T 71-2018).

More detailed techniques involved in shallow seismic exploration including equipment selection, data acquisition, seismic wave excitation, data processing, accuracy requirements and section material interpretation are suggested for referring to "*Surveying and prospecting of active fault*" (GB/T 36072-2018).

More technical requirements including preliminary preparation, site selection, topographic mapping, trench excavation, trench record preparation, trench interpretation and paleoseismic identification, sampling and dating and results in paleoseismic trench work are suggested for referring to "*Surveying and prospecting of active fault*" (GB/T 36072-2018) and "*Active fault survey—Paleoseismic trenching*" (DB/T 81-2020).

3.4 Procedures

The seismic and geological survey mainly includes four working stages: indoor preparation, field survey, data analysis and summary, results and data archiving. The specific process is shown in Fig. 3.1.

1. Indoor preparation stage

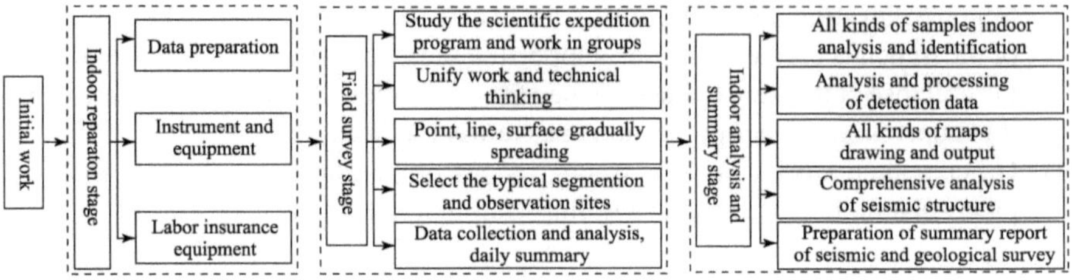

Fig. 3.1 Flow chart of seismic and geological survey

1) Data preparation.

(1) The known parameters of the earthquake should be collected.
(2) Seismic and geological data which is related to the earthquake area should be collected, including regional geological survey reports and hydrogeological survey reports (1:200,000).
(3) Different scale geological maps and topographic maps which is related to the earthquake area should be collected, especially 1:10,000–1:50,000 topographic maps and geological maps used in field mapping.
(4) High-resolution satellite and aerial images related to the earthquake area should be collected and preliminarily interpreted.
(5) The work content, focus and route of field survey are made.
(6) It is suggested that qualified departments should collect and sort out data and materials in danger areas and even nationwide on a daily basis, so that post-earthquake scientific survey can quickly obtain relevant data and materials.

2) Equipment and labor insurance equipment.

(1) All kinds of field observation instruments and tools mainly include hand-held GNSS, differential GNSS, geological compass, laser rangefinder, geological hammer, ice axe, 3D laser scanner, UAV and airborne LiDAR.
(2) All kinds of field work equipment and insurance equipment should be prepared.

2. Field survey stage

According to the general requirements of the scientific expedition program, this guide should be applied to carry out the corresponding seismic and geological survey.

(1) Due to the difficult working conditions, urgent time and heavy tasks, field work should be divided into several groups, which should strengthen communication and carry out in accordance with the unified working methods and technical solution.
(2) Field work should be gradually carried out and form a system in the way of point, line and surface.
(3) The work of the survey sites should be carried out steadily, and the data should be as rich and accurately recorded as possible.

(4) On the basis of the field survey, the typical sites and phenomena should be selected as the short-term and long-term priority.
(5) We should carefully organize and comprehensively analyze the data of field work and make a daily summary, so as to prepare the summary report of field work in time after the field work.

3. Indoor analysis and summary stage

(1) Indoor analysis and identification of dating samples, rock samples and geochemical samples.
(2) Analysis and processing of various exploration data mainly including geophysical and geochemical exploration.
(3) Remote sensing interpretation map, geological and geomorphologic plane maps, geological and geomorphologic section maps, earthquake surface rupture distribution maps, seismic structure map, fault section map, trench section map, geological hazard distribution map.
(4) Comprehensive analysis of seismic structures.
(5) Summary report of seismic and geological survey.

3.5 Results and Data Archiving

1. Content and form of achievement

The achievements of seismic and geological survey mainly include the geometry and kinematic characteristics of earthquake surface rupture zone, the distribution characteristics of co-seismic surface displacement and seismic and geological hazards, the seismic structural environment and the determination of seismic structures. The illustrated survey reports and related attachments are submitted after the completion of the survey work.

2. Main contents of the summary report of seismic and geological survey

The summary report of the seismic and geological survey includes but not limited to the following main contents.

1) Introduction.

The introduction includes work objectives, work content and workload, survey methods and work deployment, the list of working group members and division.

2) Seismic structural background of the earthquake area.

(1) Regional geological structure setting.
(2) Regional topographic and geomorphic characteristics.
(3) Characteristics of regional neotectonic movement.
(4) Regional active fault distribution.

(5) Regional deep structures and geodynamic environment.

3) Mapping of earthquake surface rupture zones.

(1) Survey of the location, shape and scale of the surface rupture zones and mapping of geometric distribution.
(2) Development characteristics of surface rupture zone and local map of typical rupture characteristics.
(3) Survey of associated structures in the surface rupture zones.
(4) Survey and determination of the maximum co-seismic displacement.
(5) Measurement and statistics of rupture zone displacement and analysis of displacement distribution characteristics.
(6) Comprehensive analysis of geometric and kinematic characteristics of earthquake surface rupture.

4) Seismic structural environment.

(1) Quantitative study on the late Quaternary activity of seismic faults. The acquisition process of quantitative parameters including kinematic characteristics, attitude, scale, segmentation, paleoearthquake recurrence models and intervals and slip rates of the seismic faults are discussed.
(2) The late Quaternary activity survey of other related faults in the earthquake area. The published data of active faults survey and mapping are collected. Furthermore, the faults with weak study and unknown late Quaternary activity should be further surveyed and the fault activity for should be identified.
(3) Survey results of other active structures and combination characteristics of active structures in the earthquake area. The properties, scale, periods and age of latest activity of active folds, active basins and active volcanos should be discussed, and the control and influence of these active structures on earthquake also need to analyze.
(4) Mapping of seismic structure map of earthquake area. The characteristics of seismic structure environment should be analyzed and discussed.

5) Surface deformation zones and seismic and geological hazards.

(1) Distribution characteristics of various surface deformation.
(2) Direct seismic and geological hazards, including surface dislocation, earthquake collapse, landslides, liquefaction and subsidence.
(3) Secondary seismic and geological hazards, mainly including dammed lakes, fire disasters, disclosure of dangerous goods and plague.
(4) Characteristics of historical earthquakes and geological hazards of this earthquake. Hazards distribution map should be attached.

6) Comprehensive analysis of seismic structures.

(1) Temporal and spatial distribution characteristics of earthquakes in earthquake area.

(2) Intensity distribution, dislocation and modern stress field characteristics of this earthquake.
(3) Long-term activity of seismic structures.
(4) Deep structure characteristics of seismic structures.
(5) Comprehensive analysis of seismic structures.

7) Problems and deficiencies.

Comprehensive summary of the seismic and geological work should be made and then compare the results of the work with the expected targets, put forward the existing problems and deficiencies and provide reference for the future scientific expedition.

3. Raw data archiving

Raw data of seismic and geological survey work should be submitted but not limited to the following contents.

(1) Data of fundamental geology, mapping and geography related to seismic and geological survey.
(2) All kinds of raw measurement (mapping) data obtained in field survey, mainly including three-dimensional coordinates of survey points, profile data, three-dimensional terrain data, UAV aerial photography data and displacement data.
(3) Field books and various data and plane maps and section maps at various scales in the geological survey.
(4) All kinds of raw measurement data of geophysical and geochemical explorations.
(5) Overall photos, partial photos, grid photos and splicing photos of trench.
(6) Field sampling photos of age samples and test reports.
(7) Construction photos of shallow seismic exploration, drilling core photos, record from drill core, borehole histogram and section map.
(8) All kinds of field geology and landform, photos of geological hazard phenomenon, remote sensing images and caption.
(9) The archiving of surface rupture zones mapping and active faults mapping should refer to relevant requirements of "*Active fault survey—Inspection of database*" (DB/T 83-2020).
(10) Confidential data should be archived and managed according to the relevant state confidentiality and management measures.

Chapter 4
Seismic Disaster Remote Sensing Comparative Investigation and Acquisition

4.1 Objectives

By collecting post-earthquake remote sensing images of typical cities, towns and villages in different seismic intensity zones, we will analyze buildings (structures), lifeline engineering, infrastructures, earthquake damage degree and contributing factors of post-earthquake area on the image, investigate on-site and verify secondary geological hazards. Inadequacy of ground-based seismic surveys will slow down the research progress such as identification of seismic hazards, quantitative assessment, seismic intensity assessment and damage assessment. To solve this, we can use the low altitude UAV data to obtain fine-grained 3D ground data and complete the remote sensing images of earthquake-induced disaster contribution to further research.

4.2 Contents

1. Remote sensing field comparison survey of earthquake damage

(1) On-site survey and verification of buildings (structures), lifeline engineering and infrastructures.

Detailed investigation of building damage information is carried out in typical cities, towns and villages. These survey areas are selected from remote sensing images by considering factors such as seismic intensity, terrain topography and geological formations. The verification work is to see the damage characteristics, structural type, usage, destruction reason of buildings and lifeline engineering on-site and compare to the identification results from remote sensing images after earthquake.

(2) On-site survey and verification of earthquake-induced secondary geological hazards.

Institute of Earthquake Forecasting, China Earthquake Administration,
A Guidebook to Earthquake Scientific Investigation,
https://doi.org/10.1007/978-981-96-0418-0_4

Earthquakes will cause a series of minor disasters, like collapse, landslide, weir, mudflow, fracture, soil liquefaction and soft soil seismic trap. Therefore, we need both field research and remote sensing image interpretation to do quantitative research of the scale and distribution of these phenomena.

2. UAV data acquisition of typical earthquake damage scenario

According to the post-earthquake emergency response phase UAV data and remote sensing images of earthquake hazards, we will first determine the area and resolution of UAV acquisition in the disaster area and develop a schedule for on-site UAV data acquisition. Then obtain and process optical, radar and LIDAR data of typical areas where the earthquake hit. Finally, mosaic process the UAV data, extracts and analyze fine features of buildings, road traffic, and prepare image maps of earthquake damage distribution for each disaster element.

3. Fine 3D data ground acquisition of typical object

On the basis of the remote sensing comparison survey, we pick some typical damage targets such as buildings, structures, major engineering facilities, road traffic and its auxiliary facilities and landslides in the disaster area and then use high-precision ground-based 3D observation instruments such as ground-based LIDAR and panoramic cameras to obtain fine 3D data of typical targets post-earthquake scenes. Sequentially, we can construct 3D visualization models and quantitative 3D earthquake damage scene images of typical targets and analyze their characteristics and earthquake-induced-hazard mechanisms.

4. Comprehensive analysis of earthquake damage in disaster areas and preparation of remote sensing earthquake damage distribution maps

Based on the above research, we interpret remote sensing images and then analyze the distribution of buildings, lifeline engineering, infrastructure damage and secondary geological hazards in the disaster area comprehensively. After that, we rectify the remote sensing identification results according to the field survey results and establish the correspondence between remote sensing damage and ground survey. Thus, for the assessment points without a ground survey, we can perform a quantitative analysis of the damage degree based on the newly established quantitative model, and compile the distribution map of earthquake damage in the disaster area.

4.3 Methods and Basic Requirements

1. Remote sensing field comparison survey of earthquake damage

1) Methods.

According to the reference map of the remote sensing comparison survey, select a certain area around the survey site, and we will record, measure and take photos of all buildings (structures), lifeline engineering and infrastructures in this area to obtain their structure type, height, construction age, usage, whether they are fortified, seismic intensity, casualties, damage pattern/area/scale, site conditions, etc. Meanwhile, we will investigate the overall number of buildings (structures), the proportion of structural types, the number of people, the damage and other basic information about the survey site. For earthquake minor disasters such as collapse, landslide, weir, mudflow, fracture, soil liquefaction and soft soil seismic trap, we record their location, geometry and distribution characteristics, measure their scale and displacement and take photos.

2) Basic requirements.

(1) Taking into account factors such as seismic intensity, topography, disaster distribution by remote sensing interpretation and seismic hazard anomalies, we single out a certain number of urban and township neighborhoods and natural villages in various environments within different intensity zones, and the number and distribution of sampling points of housing buildings in the disaster area are referred to GB/T 18208.3-2011 (*Post-earthquake field works—Part 3: Code for field survey*).
(2) Sample survey point distribution should cover the hardest-hit areas, survey points should represent different degrees of damage, and different types of structures, and each survey point on the building sampling points should be no less than 100 buildings.
(3) According to GB/T 18208.3-2011, investigate the type and damage degree of construction (structure) buildings, lifeline engineering and infrastructure one by one in the sample area.
(4) Locate and mark the construction (structure) buildings, lifeline engineering, and infrastructure on the image map.

2. UAV data acquisition of typical earthquake damage scenario

1) Methods.

According to the data acquisition area terrain conditions and disaster investigation needs, determine the regional scope of the flight, choose fixed, rotary, vertical takeoff and landing and other models with optical, radar, LiDAR, tilt photography and other sensors, determine the image resolution, overlap, altitude, data accuracy and other data acquisition requirements, develop a flight plan, the survey area for UAV data acquisition.

2) Basic requirements.

(1) We recommend that the image resolution of buildings (structures), lifeline projects and infrastructure should be no less than 0.2 m, and the overlap rate should be no less than 50%.

(2) Before obtaining the UAV data, read the Civil Aviation Administration of China Civil UAS Operation Management Regulations carefully, and obey it strictly.
(3) Make an elaborated flight plan before data acquisition, and according to the local meteorological conditions, select the right flight time, which is beneficial to the image data quality.

3. Fine 3D data ground acquisition of typical object

1) Methods.

Depending on ground-based high-resolution 3D data acquisition technologies such as ground-based 3D laser scanners and panoramic cameras, obtain high-precision 3D cloud data and high-definition real-scene photos of buildings, lifeline engineering and infrastructure and their seismic impact factors in the typical earthquake-hit area.

2) Basic requirements.

(1) The target selection of ground 3D data should be representative and typical, and try to pick real scenes with different structural types, different degrees of damage of buildings (structures), lifeline engineering and infrastructures for 3D data acquisition and measurement analysis.
(2) While collecting 3D point cloud data, it is necessary to take images with spatial positions. The point cloud density and image clarity are as high as possible when the observation conditions allow.
(3) The ground acquisition base station layout should consider the integration and integrity of the observation target.

4. Comprehensive analysis of earthquake damage in disaster areas and preparation of remote sensing earthquake damage distribution maps

1) Methods.

(1) Data preprocessing.

Preprocessing such as geometric exact correction, fusion, mosaic and registration of remote sensing images before and after earthquakes. Splicing multiple unmanned aerial vehicle images and LiDAR data obtained on-site to produce an orthophoto map, 3D digital surface model (DSM), 3D elevation model (DEM) and other products; using ground-based radar to collect high-precision 3D point clouds and panoramic high-precision photos, and stitching data from multiple base stations to generate 3D fine measurement data for each observation target.

(2) Earthquake damage remote sensing extraction.

According to on-site survey and remote sensing comparison survey information, combined with preprocessed image data, the image characteristics of buildings (structures), lifeline projects and infrastructure of different structural types and different degrees of damage were analyzed. Establish its interpretation marks on images of different resolutions and different sensors, and on this basis, extract earthquake damage information by visual interpretation, automatic or semi-automatic extraction.

Visual interpretation should be based on the remote sensing interpretation signs (*Remote sensing assessment of earthquake hazards—Earthquake geological hazards* (DB/T 74-2018), *Remote sensing assessment of earthquake hazards building damage* (DB/T 75-2018), *Remote sensing assessment of earthquake hazards highway earthquake damage* (DB/T 76-2018)) of buildings, highway earthquake damage and secondary geological disasters in the series of standards for remote sensing assessment of earthquake sites, and human–computer interaction should be used to determine their structural types, damage levels and spatial distribution.

According to the on-site investigation data, selecting earthquake damage samples such as buildings and road traffic and methods such as supervised or unsupervised classification, object-oriented classification and deep learning are used. If there are images before the earthquake, the change detection method is used to automatically or semi-automatically identify the spatial location, structure type and earthquake damage degree of the building, and estimate its area. Based on the 3D seismic damage data obtained by LiDAR, oblique photography, etc., 3D modeling technology is used to construct a 3D model of the earthquake damage scene and 3D information extraction and 3D visualization technology are used to finely measure the damage degree of buildings (structures).

2) Basic requirements.

(1) The images collected after the earthquake should include airborne (including UAV) and space-borne high spatial resolution images before and after the earthquake, and cover the hardest-hit areas as much as possible.
(2) For the lowest spatial resolution of image data, earthquake intensity, disaster extraction factors should be considered comprehensively. Considering factors such as image and extraction methods, available images, etc., sub-meter optical images are preferred.
(3) The image format should meet the format requirements of remote sensing image processing software and should include image files and data descriptions.
(4) When there are multi-temporal remote sensing images in earthquake-stricken areas, one temporal image should be used as a benchmark to register other temporal images.

4.4 Procedures

The remote sensing comparison survey and disaster situation acquisition includes four stages of work: indoor preparation, field site investigation, data analysis and summary, and results and data archiving, as shown in Fig. 4.1.

1. Indoor preparation

1) Remote sensing data system processing in disaster areas.

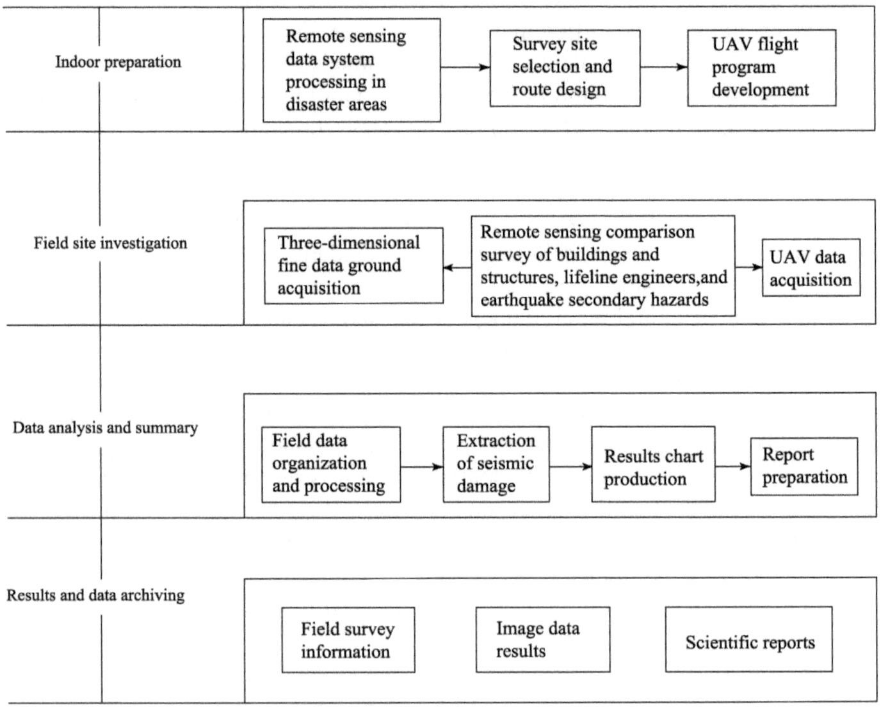

Fig. 4.1 Flow chart of remote sensing comparison survey of earthquake damage

The remote sensing images of different types, different time phases and different resolutions of the disaster area are collected and systematically preprocessed, such as calibration, alignment and mosaic, to establish the accurately calibrated remote sensing images of the disaster area, which provide the basis for the detailed extraction of the earthquake damage in the disaster area.

2) Design of examination site selection.

According to the scope of the disaster area, remote sensing data acquisition, remote sensing damage distribution and earthquake emergency response phase site investigation, etc. It should choose different types of structures, different degrees of damage in the typical city and township neighborhoods, and natural villages as field scientific investigation points, design the corresponding field investigation routes, and develop a detailed field investigation plan.

3) Formulate a UVA flight plan.

Combined with the post-earthquake remote sensing damage characteristics and image data, determine the expected operating range of UAV data acquisition, determine the technical parameters such as sensor type, image resolution, accuracy, overlap, etc. according to the observation target task, carry out zoning and route planning design for each observation area and preliminarily select the landing and takeoff points in combination with high-resolution images. According to the observation needs, choose to carry out a ground measurement and control, and carry out the corresponding base station layout design.

4) Instrument and material preparation.

(1) According to the survey plan, made reference images and survey forms for remote sensing comparison survey, and prepared survey equipment such as laser rangefinders, GNSS instruments and cameras.
(2) Preflight test of the drone, checking and equipping with adequate airwaves, parachutes, paddles, accessories, etc.
(3) Prepare and inspect ground acquisition equipment such as 3D laser scanners, panoramic cameras, etc.

2. Field investigation

1) Remote sensing comparison survey.

According to the field scientific research program, the field verification work of remote sensing images of seismic damage was carried out at the survey site. GNSS spatial positioning, structural type, structural form, degree of earthquake damage, site conditions, disaster-causing factors and remote sensing image representation of earthquake damage are carried out for the buildings (structures), lifeline projects and major engineering earthquake damages at each survey site. For surveys and inquiries, mark one by one on the survey reference image map, fill in the survey form in detail, take photos from long-range, close-range, local and other angles, locate by GNSS, measure cracks, displacements, etc. Select typical buildings (structures) and secondary disasters for high-precision quantitative measurement on the 3D ground.

2) UAV data acquisition for typical areas.

According to the flight plan of the UAV, the field survey of the observation area is carried out to determine the landing and takeoff points of the aircraft, to plan the route and to prepare for the flight by checking the power supply, battery, sensors and storage system of the aircraft. When there are multiple sorties, the original data should be backed up in time. When there are local defects in the quality of the original data, it is necessary to make up the flight or re-fly.

3. In-house analysis and summary phase

1) Field data organization and processing.

The images and tables of the site survey were digitized, and the photographs taken on-site were organized. Production of orthophotos, DSM and DEM based on UAV images acquired in the field and the 3D earthquake damage model is built based on the point cloud or panoramic images collected on the ground.

2) Seismic Damage Extraction and Results Production.

Based on the processing of aerial and satellite remote sensing data and on-site investigation data in the earthquake area, we extracted detailed information on the damage to buildings, lifeline projects and megaprojects in the disaster area, calculated the seismic damage index and earthquake intensity of the residents and extracted various secondary hazards. Then produce remote sensing image maps at different scales and prepare the remote sensing comparison survey for earthquake hazards and disaster acquisition reports.

4.5 Results and Data Archiving

1. Content and form of the results

The main results include seismic damage site investigation data of buildings, lifeline projects and major projects earthquake damage site investigation information, typical earthquake damage scenario UAV data, typical damage object ground fine 3D data, remote sensing of earthquake distribution traits and quantitative damage assessment models, etc. After the survey, the results were submitted in the form of graphs, photos, video data and survey reports.

2. The main content of the summary report

The report should include but is not limited to, the following key elements.

1) Preface.

(1) Basic overview of the earthquake, including basic parameters of the earthquake, basic conditions of the disaster area, seismotectonic environment and physical geography of the earthquake area.
(2) Remote sensing comparison survey and drone acquisition work objectives, work content and workload, survey methods and work deployment, working group personnel list and division of labor.

2) Earthquake disaster remote sensing data collection and processing.

(1) Post-earthquake remote sensing data acquisition and coverage.
(2) Remote sensing data processing.
(3) Remote sensing rapid assessment of earthquake disasters in the post-earthquake emergency response phase.

3) UAV data acquisition and processing in disaster areas.

(1) UAV data acquisition, including area coverage, resolution, etc.
(2) UAV data processing methods and results.

4) Remote sensing survey and assessment of earthquake damage to buildings in disaster areas.

(1) On-site investigation of the building (structure) situation, including the structure type, distribution characteristics, etc.
(2) Remote sensing image characteristics of typical seismic damage of buildings (structures).
(3) Remote sensing extraction methods and results of seismic damage of buildings (structures).
(4) Remote sensing earthquake damage index in disaster areas and establishment of quantitative conversion model for actual earthquake damage.
(5) Integrated remote sensing assessment of earthquake intensity.

5) Remote sensing comparison survey of lifeline projects and megaprojects in disaster areas for earthquake damage.

(1) Lifeline projects and major projects situation.
(2) Seismic remote sensing image features of lifeline projects and megaproject.
(3) Remote sensing comparison survey and seismic damage distribution of lifeline projects and megaprojects.

6) Remote sensing interpretation of seismic and geological hazards in disaster areas.

(1) Remote sensing image characteristics of typical earthquake and geological hazards.
(2) Remote sensing interpretation and assessment of seismic and geological hazards.

7) Conclusion and problems.

(1) Distribution characteristics of seismic hazard and seismic intensity.
(2) Earthquake mitigation measures and recommendations.
(3) Problems and shortcomings.

3. Raw data aggregate submission

The original data submitted for the seismic disaster remote sensing comparison survey and disaster acquisition work include but are not limited to the following content.

(1) Various information and data of remote sensing comparison field survey of seismic hazards, including mapping the location of survey point Reference image maps, survey sheets, photos, GNSS positioning information, etc.
(2) UAV-acquired data in typical areas, including orthophoto, DSM, DEM, etc.
(3) Highly accurate 3D scene data on the ground, including 3D point clouds, 3D stereo models, etc.
(4) Distribution of earthquake damage images at different scales.

(5) Classified information is archived and managed in accordance with relevant national secrecy management methods.

Chapter 5
Crustal Deformation Observation

5.1 Objectives

Conduct GNSS campaign station resurveys in the epicentral zone and surrounding areas. If there is a precise leveling network in the epicenter, resurvey the leveling network. Acquire co-seismic displacement field derived from GNSS, InSAR and level re-measurement data. Construct new temporary continuous GNSS stations and conduct GNSS observations aside from existing GNSS continuous and campaign stations. Finally, Crustal deformation and fault kinematics in the epicentral zone and its surrounding faults are to be analyzed to benefit the assessment of regional seismic activity.

5.2 Contents

1. GNSS campaign observation and leveling resurvey

According to the characteristics of the seismic tectonics, for magnitude 6–7 earthquakes, we should select GNSS campaign stations within 100 km of the epicenter to carry out post-seismic re-measurement and to remeasure the existing level lines. For magnitude 7–8 earthquakes, we should select GNSS campaign stations within 150–200 km of the epicenter to carry out post-seismic resurvey, and also to carry out resurvey of existing leveling lines in the epicentral area.

2. Extraction of co-seismic deformation fields

Using the observation data of continuous and campaign GNSS stations before and after the earthquake, the co-seismic horizontal displacement field is calculated by precise positioning solution. The co-seismic vertical displacement field is given based on the leveling re-measurement data of the epicenter and its surroundings. Based on

Institute of Earthquake Forecasting, China Earthquake Administration,
A Guidebook to Earthquake Scientific Investigation,
https://doi.org/10.1007/978-981-96-0418-0_5

repeat-orbit SAR satellite observation data, near-field co-seismic LOS deformation field or three-dimensional deformation field is given. Finally, high-resolution three-dimensional co-seismic deformation field is given by combining GNSS, leveling and SAR results.

3. Co-seismic rupture model inversion and co-seismic Coulomb stress change analysis

Use co-seismic displacement field obtained by geodetic observation to invert for source rupture model with respect to dataset weight. Coulomb stress changes at different depths of the seismogenic fault surface should also be calculated. Geometric parameters and kinematics of active faults around the epicenter should be collected and collated, and the Coulomb stress changes caused by the source rupture on seismogenic faults in the epicenter zone are needed to be calculated.

4. Analysis of post-seismic deformation characteristics based on GNSS observations

Some temporary GNSS continuous stations should be built in and around epicenter. Observation station distribution is determined according to the type of seismogenic fault and the spatial distribution of fault rupture. The basic principle of the station layout is to build relatively dense GNSS stations at both ends of rupture belt to make inversion for the slip rupture model better constrained. Through continuous observations, the coordinate time series of the new GNSS stations should be continuously analyzed to extract post-seismic deformation characteristics. In addition, the pre- and post-seismic GNSS observations should be employed to analyze the characteristics of regional crust deformation, response of surrounding faults to the earthquake and to assess earthquake potential in the epicenter and its surrounding regions.

5. Analysis of post-seismic deformation characteristics based on SAR satellite observations

Use ascending and descending data time series of SAR satellites to generate interferogram time series by PS-InSAR method or small-baseline array method. Post-seismic deformation sequence and deformation field should be extracted using DEM error processing, atmospheric error and orbit error correction. Based on the after-slip, viscoelastic relaxation and pore rebound model, the post-seismic deformation mechanism should be analyzed. The loading or unloading effects of co-seismic and post-seismic deformation on faults around the epicenter zone should be analyzed in conjunction with co-seismic deformation.

5.3 Methods and Basic Requirements

1. GNSS campaign observation, leveling line resurvey

1) Working methods.

High-precision GNSS resurvey and leveling survey line resurvey should be carried out in the epicenter zone and its surrounding areas.

2) Essential requirements.

The basic process of GNSS campaign observation and leveling line resurvey are shown in Fig. 5.1.

(1) Basic requirements for GNSS campaign observation.

Before each campaign, the operation implementation plan should be prepared first, then submitted to the organization in charge of the project for approval, and then issued to each operational group. The implementation plan should include task source, content to do, and an overview of the area to be surveyed. The design of

Fig. 5.1 The basic process of GNSS campaign observation and leveling line resurvey

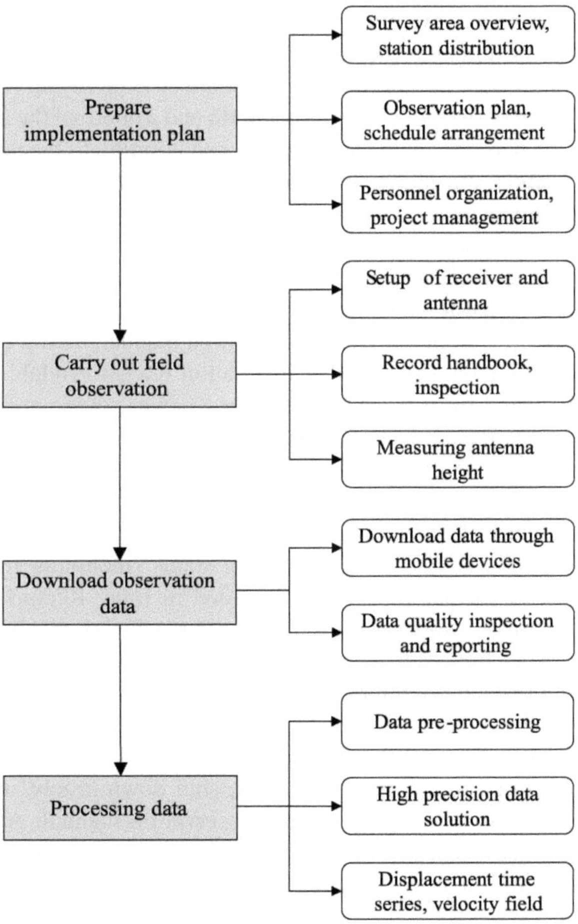

the observation program includes what principles to be based on, instrumentation and its inspection, GNSS site distribution map, progress arrangement and schedule of the observational plan, data recording and data submission, personnel organization and its management, etc. Observations at each station should be completed within 7 days for the specified effective observation period, with effective observation sessions ≥ 3. An observation session is defined as 0:15–23:45 UTC (8:15–7:45 BST) each day, and 23:45–0:15 UTC is the time for data download.

Antenna setup: Connect the antenna firmly, align it to mark embedded into the monument, and level. Rotate the antenna so that its directional line points straight north (use a compass, and add magnetic declination correction), directional error should be less than 5° and lock the antenna. Measure the antenna height and record the measurement position and method in detail in the handbook. Antenna height is the vertical distance from the antenna reference surface (bottom of the amplifier) to the surface of the observational pillar. The antenna height measurement should be taken at three positions of the antenna reference plane at an angle of 120° to each other, and the average value of the three positions should be taken as antenna height data. The reading and average value should be accurate to 1 mm. For each observation session before and after the measurement of antenna height, if the difference is greater than 3 mm, the observation session should be resurveyed. After the antenna is setup, take a photo of the antenna pointing north and a photo of the antenna's height measurement. The photos should show the antenna pointing to the north, the antenna measurement method, measurement location, measurement value, point number, and date of the photo taken.

Receiver setup: According to the length of the antenna cable, place the instrument in a suitable place smoothly; connect the antenna to the receiver with the cable and fasten it. Connect the receiver to the power source according to the procedure specified in the instrument operation manual, taking care not to reverse positive and negative battery poles. If the station has an available AC power source, it can be powered by both an AC-regulated power source and a battery. Make sure that all cables are connected correctly before turning on the receiver power switch.

Field observation: After the observation starts and the receiver starts recording data, an observer should check the receiver frequently, including parameter settings, number of receiving satellites, satellite number, the signal-to-noise ratio of each channel, and memory recording status (including remaining memory). Problems should be dealt with and recorded in time. Record day and weather conditions. Record weather conditions according to the levels: clear, less cloudy, cloudy, heavily cloudy, light rain, moderate rain, heavy rain, sleet, snow, heavy snow, wind (direction of the wind, the strength of the wind, such as southerly 5°). Observers should operate the instrument carefully to prevent it from touching the antenna or blocking satellite signals. Pay attention to equipment to prevent exposure to the sun and rain. In the event of strong thunderstorm weather, shut down timely, and remove the antenna cable, make detailed records in the observation manual. At the end of each observation period, measure post-measurement antenna height and take a record of it, and then turn off the power after confirmation, and prepare to download observational data.

Data download: Use the software provided by the receiver. Bad data blocks should be monitored during data download. After completing the data download, check it against the instructions. If not fully downloaded (< 100%) or a bad block of data appears, reset the communication parameters and download again.

(2) The basic requirements of leveling observation.

Leveling measurement should be conducted according to the relevant requirements of "*National first and second degrees of leveling specification*" (GB/T 12897-2006), "*Compilation of regional leveling technical documents*" (National Seismological Bureau, 1996), and "*Digital leveling technical specification for seismic terrain deformation*" (DB/T 5-2003). Records shall be made using an electronic logger (for special areas not suitable for electronic recording, manual recording may be used upon approval) and the recording procedure shall be performed using software approved by the project manager. If there is no special case, the round-trip measurement of a section should be observed by the same observer. The field data acquisition of leveling is recorded by computer, and a unified electronic recording program is applied. The integrity and authenticity of data must be guaranteed. Leveling observation field results in accordance with the "*National first and second degrees of level measurement specification*" (GB/T 12897-2006). After Leveling work is completed, it should be assessed in accordance with the "*Surveying and mapping results quality inspection and acceptance*" (GB/T 24356-2009).

2. Acquisition of co-seismic displacement field

1) Working methods.

Co-seismic horizontal displacements should be derived from existing GNSS continuous station data in the epicenter and surrounding areas, combined with pre-seismic and post-seismic GNSS campaign station data. Co-seismic vertical displacements should be acquired based on the leveling surveys before and after the earthquake. Co-seismic LOS deformation field and azimuth deformation field should be calculated based on pre-seismic and post-seismic interferometry of SAR satellite.

2) Essential requirements.

GNSS data collection: Collect raw observational data from all GNSS stations within 200 km of the epicentral area 3–7 days before and after the earthquake, respectively, and pre-seismic coordinate time series. The raw data from GNSS stations of the China Crustal Movement Observation Network, the GNSS observation network established by each organization, and the IGS reference frame network 3–7 days before and after the earthquake, respectively, should be collected. Collect the IGS standard satellite ephemeris and other auxiliary files in the same period. If the satellite ephemeris is not released in time after the earthquake, it can be replaced by the fast ephemeris products provided by IGS.

Other data collection: Collect high-precision and high-resolution DEM data in and around the epicenter, such as SRTM, pre-seismic and post-seismic SAR satellite data, and precise orbits data.

Data Processing: Based on GNSS data of the epicenter and surrounding areas, a regional GNSS observation, and a subnet for data processing should be formulated. The raw observation data of the regional network 3–7 days before and after the earthquake, respectively, should be solved by using high-precision GNSS data processing software, and a daily solution normal equation of GNSS coordinates can be obtained. Based on "China Crustal Movement Observation Network", the GNSS observation network established by each organization, the IGS reference frame network and other GNSS stations, a global GNSS observation subnet for data processing should be formulated, ensuring coincidence of more than three GNSS stations between the global network and the regional network. The raw observation data of the global network 3–7 days before and after the earthquake are solved to obtain a single-day solution normal equation of the global network GNSS coordinates. The regional network and the global network GNSS station coordinate single-day solution normal equations are bound together to obtain fused GNSS station coordinate single-day solution normal equations, and coordinates of IGS stations within the ITRF reference frame are used as constraints. The GNSS network is bound to the ITRF reference frame to obtain the coordinate time series of the GNSS stations in the epicenter and its adjacent areas 3–7 days before and after the earthquake, respectively. The horizontal accuracy of the single-day solution coordinates of the GNSS stations is required to be less than 2 mm, and the vertical accuracy is less than 5 mm. The horizontal uncertainties of the GNSS station coordinate time series are less than 3 mm, and the vertical uncertainties are less than 10 mm.

Based on the pre-seismic and post-seismic SAR data, the interferometry are used to eliminate DEM phase contribution, suppress the influence of atmospheric error, orbit error and DEM error, and solve the co-seismic deformation field. The deformation accuracy should reach at least a centimeter level. Range and azimuth movement fields should be calculated through the pixel matching method, and deformation accuracy should reach at least 1/20–1/40 pixel size.

Displacement calculation: For the continuous GNSS stations, their mean values should be calculated for their coordinates 3–7 days before and after the earthquake, respectively, and the three-dimensional coordinates of the GNSS continuous stations before and after the earthquake are obtained respectively. For the GNSS campaign stations, based on the time series of the coordinates before the earthquake, the velocity of the campaign stations before the earthquake is fitted, and three-dimensional coordinates of the campaign stations before the earthquake are obtained by the interpolation method. The mean values of the station coordinates after the earthquake are calculated to obtain their three-dimensional coordinates after the earthquake. The three-dimensional coordinate differences and their errors before and after the earthquake should be solved to obtain the co-seismic three-dimensional displacement field and its error.

Based on the LOS co-seismic deformation field extracted by ascending and descending SAR satellite interferometry and the azimuth deformation field obtained by the pixel matching method, the co-seismic three-dimensional displacement field should be also solved.

Co-seismic rupture modeling: The geometric parameters and slip distribution of rupture faults should be inverted, based on the constraint of co-seismic deformation fields, and the error of each parameter should be also estimated.

3. Post-seismic deformation analysis based on GNSS data

1) Working methods.

For earthquakes occurring in an area with weak monitoring capabilities, GNSS observations should be intensified in and around the epicenter. Two ways can be used to intensify observations: construct new continuous GNSS stations to carry out continuous observation for more than half a year; construct new campaign GNSS stations and observe for at least three sessions. For earthquakes occurring in an area with good monitoring capabilities, a post-seismic deformation monitoring scheme can be determined according to the research objectives, especially for the new continuous station and campaign station in the affected area of the earthquake. The post-seismic impact of the earthquake should be analyzed, referring to the characteristics of background crustal deformation.

2) Essential requirements.

(1) For new GNSS stations, their constructions and observations should meet the relevant requirements of the "Crustal movement observation network of China", as stated in "*Technical specification for crustal movement monitoring*" (China Environmental Science Press, 2013).
(2) Analysis of post-seismic deformation. Focus on the co-seismic and post-seismic deformation characteristics of the earthquake, especially the latest movement of the major faults in the tectonic area with significant deformation, and provide the basis for the determination of regional earthquake potential.

4. Analysis of post-seismic deformation based on SAR satellite observations

1) Working methods.

Post-seismic deformation field extraction: Co-seismic deformation field, post-seismic deformation sequence and cumulative deformation are extracted by using repeat-orbit SAR satellite observations.

Post-seismic deformation mechanism inversion: Based on post-seismic deformation times series and accumulated deformation field constraint, through the three kinds of post-seismic deformation model: afterslip, viscoelastic relaxation and pore rebound, invert for and analyze post-seismic deformation, and determine which mechanisms the post-seismic deformation agree with.

Analysis of fault stress variation and earthquake potential: The Coulomb stress variation of active faults around the epicenter should be calculated by using the fault slip model of the co-seismic and post-seismic deformation field.

2) Essential requirements.

SAR satellite interferometry: Multiple SAR satellite observations should be available for the epicenter zone, and preferably with ascending and descending observations, with repeat visit intervals ≤ 12 days.

Less vegetation coverage, flat terrain, and good coherence of interferences are preferred in the epicentral area.

5.4 Procedures

The observation of crustal deformation can be divided into three stages: preparation, field investigation, and analysis and summary. The specific workflow is shown in Fig. 5.2.

1. Preparation stage

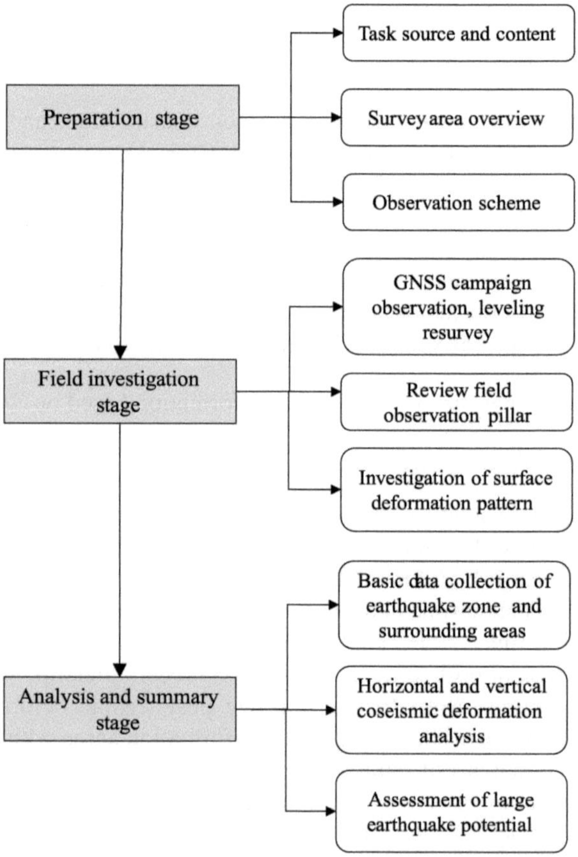

Fig. 5.2 The workflow of crustal deformation observation

Before each campaign observation, each operating group should prepare an implementation plan, submit it to the project management organization for approval, and issue it to each operating group. Implementation programs include:

(1) Task source and content.
(2) Survey area overview.
(3) Observation scheme design, including design principle, instruments and their inspection, GNSS observation stations distribution map, observation plan and schedule, data recording and sharing.

2. Field investigation stage

(1) GNSS campaign stations and leveling route resurvey in and around the epicenter. Specific requirements refer to the basic requirements of the aforementioned campaign GNSS and leveling route resurvey.
(2) After the initial co-seismic displacement is given, it is necessary to review the reliability of co-seismic displacement of significant earthquakes in the field. The work mainly includes the investigation of the observation environment around the observation point, the inclination of the observation pier and the working condition of the receiver.
(3) According to the results of co-seismic displacement, it is necessary to carry out field investigation on the deformation mode of the surface. The work mainly includes the rupture mode of the earthquake and the co-seismic influence area.
(4) Personnel organization and management.

3. Analysis stage

(1) Basic data collection, collation and data processing of various disciplines regarding the epicenter area and its surrounding areas.
(2) Based on all the data of GNSS continuous and campaign sites, and all the leveling data, it should be done for the analysis of horizontal and vertical co-seismic deformation caused by the earthquake.
(3) Based on the influence of co-seismic and post-seismic deformation, especially latest motion of the main faults in the tectonic area with significant deformation, post-seismic deformation should be analyzed, and observation results of other disciplines should be fully considered.
(4) Personnel organization and management.

5.5 Results and Data Archiving

1. Content and results

(1) Result content.

① The raw data and derived results of GNSS continuous station and campaign station, and leveling route survey.
② 3D co-seismic displacement field should be given based on the resurveyed results of GNSS continuous station, campaign station, and leveling route.
③ Co-seismic LOS deformation field or three-dimensional deformation field derived from InSAR measurement.
④ Based on existing observation in and around the epicenter, combined with the resurvey results, influence of the earthquake and post-seismic regional deformation should be analyzed.

(2) Result form.

① Raw observational data of GNSS, InSAR, and leveling.
② Deformation derived from GNSS, InSAR, and leveling.
③ Pictures and report text.

2. Main contents of summary report on crustal deformation observation

It mainly includes the following six aspects:

(1) Implementation scheme of crustal deformation observations.
(2) Field observation of crustal deformation.
(3) Acquisition and analysis of co-seismic displacement field.
(4) Analysis results of post-seismic deformation.
(5) Data submission.
(6) Main conclusions.

3. Raw data submission

It mainly includes:

(1) All the raw observation data of the continuous station and campaign GNSS stations.
(2) Raw satellite data and deformation field derived from InSAR data.
(3) Raw observation data for leveling routes.
(4) Pictures of field investigation.

Chapter 6
Geochemical Observation

6.1 Objectives

A network of mobile geochemical observation was set up in the epicenter and the related tectonic zone to obtain data on fluid geochemical observation in the fault zone in the measurement area, analyze the changes of geochemical characteristics of the tectonic zone before and after the earthquake, give a map of the spatial and temporal characteristics of geochemical in the region, and the characteristics of fluid release intensity of the fault zone in the area. In combination with the per-seismic observation data, the material source and the change of water–rock reaction degree in the seismic area before and after the earthquake are given, and the short-term prediction opinions of earthquake risk in the epicenter and related tectonic areas are given.

6.2 Contents

1. Fluid geochemical investigation of hot springs, hot water wells, cold springs and sand blasted bubble holes

The fluid geochemical method is used to study the relationship between the deep and shallow fluid coupling and the seismicity of active fault zones. The relationship between fluids and earthquakes has two aspects: one is the passive effect of fluids on earthquakes, that is, the dynamic change of the fluid composition caused by the process of earthquake incubation and occurrence, which is the precursor of the fluid; other is the energetic effect of fluids, that is, fluids directly participate in the process of earthquake breeding and occurrence. In summary, the main investigation contents are as follows:

Institute of Earthquake Forecasting, China Earthquake Administration,
A Guidebook to Earthquake Scientific Investigation,
https://doi.org/10.1007/978-981-96-0418-0_6

(1) The depth of source and degassing intensity of each fluid component in different segments of active fault zones in key seismic hazard areas have been calculated and analyzed by using fluid geochemistry methods for hot springs, hot water wells, cold springs, and sandblasting holes, the coupling degree of deep and shallow fluid of active faults in key seismic risk areas is analyzed, and the depth of crust cutting in different segments of the active fault zone is evaluated in space, and the characteristics of fluid geochemistry before and after earthquakes are compared in time, the characteristics of earthquake precursor anomaly of deep fluid escape in active fault zone are given, and a more comprehensive model is established to explore the effect of deep fluid migration on the breeding and occurrence of earthquakes.
(2) The investigation of fluid-rock interaction around hot springs.

① The softening effect of the fluid on the rock leads to the decrease of deformation modulus and strength of rock materials, and the softening of rock effectively promotes the deformation and destruction of rock. Therefore, it is necessary to investigate the fracture surface of the bedrock and collect samples of fault gouge.
② Water–rock interaction occurs in rock pores, which controls the growth or closure of fractures. The dissolution of fluid promotes the growth of fractures, especially the dissolution under pressure, such as stress corrosion, etc., which can greatly accelerate the growth of cracks. The sedimentation of fluid can delay the growth of cracks, and sometimes even make cracks heal. So, it is necessary to investigate the fractures of the bedrock fracture surface and collect samples of veins.

(3) Sampling investigation and water quality evaluation were carried out on the descending spring, water well and water source of waterworks in the epicenter area, and drinking suggestions were given.

2. Investigation on degassing characteristics of the earthquake rupture zone

The active fault zone is the main passage of the earth's exhaust on land. The background of He, H_2, CO_2, CH_4, Rn, and Hg concentrations in soil gas in the fault section is higher than that in non-fault zones, and they are affected by the earthquake. The closer to the epicenter of the earthquake, the higher the concentration of He, H_2, CO_2, CH_4, Rn, and Hg in the soil gas of the fault section. The gas inside the Earth is the most sensitive and migratory fluid substance in the geochemistry field, and it is the carrier that can carry the information of the deep Earth directly to the surface. The investigation of gas release intensity in seismic fracture zone after an earthquake can give the relationship between gas release intensity and rupture size, and provide data support for seismic intensity investigation. Besides, a lot of toxic gas and radioactive gas will be released from the rupture zone after the earthquake. The release intensity of gas in the rupture zone is investigated and evaluated to provide the basis for environmental evaluation for post- earthquake reconstruction.

6.3 Methods and Basic Requirements

1. Methods and basic requirements for fluid geochemical investigations of hot springs, hot water wells, cold springs and sandblasted bubble holes

1) Sample collection for field investigation.

(1) Hydrographic geochemistry sampling.

At each hot spring site, collect four bottles of hot spring water, record the weather conditions on the day of sampling and the days before, the detailed time of sampling (several points), the color, smell, turbidity, suspended matter, the situation of bubbles, in the hot spring where the location of water samples, the water depth of the underwater, the lithology of the source outcrop, the presence or absence of calcium deposits, the lithology of the high mountain area recharged by the hot spring (photographs), and records of the relevant information dictated by local people. Rubber gloves were worn during sampling (to avoid the addition of pollution sources), and the water samples were bottled in polyethylene plastic bottles, which were washed three times. The washed water was not poured into the hot spring, and before canning, filtered with a 0.45 μm microporous membrane, one bottle of each anion and cation was taken, and a double sample was taken as a backup sample.

(2) Gas geochemistry sampling.

The gas sampling device consists of a glass bottle and funnel. Before sampling, the glass bottle is filled with water under the water surface, then the glass bottle is turned upside down, make the bottle mouth down, and the bottle is checked for air bubbles, insert the corked funnel under the water into the glass bottle (note that the funnel should also be free of bubbles). Move the device to the air bubble, after the water in the bottle has been drained, remove the funnel under the water, at the same time with the bottle stopper good glass bottles, glass bottles out of the water, the bottle upside down in a wooden box transported to the laboratory. It should be noted that the glass bottle must be left with a small amount of water to ensure that the gas in the bottle will not escape or air into the bottle.

(3) Sampling of main rocks and sediments in the vicinity of fluid geochemical sampling sites.

Rocks and sediments were collected from hot springs, hot water wells, cold springs, and near sandblasting holes, stored in 300 mL wide-open polyethylene plastic bottles, and labeled with sample numbers.

2) Laboratory analysis of fluid geochemical samples.

(1) Methods of hydrogeochemistry analysis.

The pH and conductivity of the water samples were determined in the laboratory using a precision portable pH meter and conductivity meter, respectively. The concentrations of hydration ions were determined with the ion chromatograph and automatic sampler with a 0.01 mg/L limit in the laboratory. Cation components tested: Li^+, Na^+, Mg^{2+}, and Ca^{2+}; anion components tested: F^-, Cl^-, Br^-, NO_3^-, and SO_4^{2-}. The concentrations of HCO_3^- and CO_3^{2-} were titrated by potentiometric titration with 0.05 mol/L hydrochloric acid. In order to calibrate the ion chromatography, the standard sample should be measured before measuring each batch of water samples, the measurement deviation is within ± 1% and the measurement error of ion components of the water sample should be controlled within ± 5%. The ion balance is calculated based on the equation:

$$(\text{ib})\% = 100 \times (\Sigma_{\text{Cation}} - \Sigma_{\text{Anion}})/(\Sigma_{\text{Cation}} + \Sigma_{\text{Anion}})$$

The trace elements of samples were analyzed by element-type inductively coupled plasma mass spectrometer (ICP-MS). The hydrogen and oxygen isotope analysis were measured by isotope mass spectrometer with δD and $\delta^{18}O$ as the international standard, namely standard mean marine water (VSMOW). The analytical accuracy of the samples was less than $\delta^2H < \pm 1‰$ and $\delta^{18}O < \pm 0.2‰$. The SiO_2 concentration was measured by an inductively coupled plasma emission spectrometer.

(2) Gas geochemistry analysis.

The N_2, H_2, CO_2, O_2, CH_4, and Ar components of hot spring gas samples are measured by gas chromatography in the laboratory. The measurement accuracy is as follows: when the content is 1–100%, the relative standard deviation is less than 0.5%; When the content is 0.01–1%, the relative standard deviation is less than 1%; When the content is 0.001–0.01%, the relative standard deviation is less than 2%; When the content is 0.0001–0.001%, the relative standard deviation is less than 5%. The concentrations and isotopes of He, Ne, and Ar in hot spring gas were analyzed by a rare gas isotope mass spectrometer. When the *R*-value of helium isotope is above 1×10^{-7}, the test error is ± 10%, and in the range of 1×10^{-8}–1×10^{-7}, the measurement error is ± 15%. The carbon isotope ratio is analyzed by a mass spectrometer. The mass spectrometer consists of gas chromatography, combustion/converter, interface, and mass spectrometer. The stable carbon isotope composition is expressed by $\delta^{13}C$. The accuracy of $^{13}C/^{12}C$ is 0.6‰.

(3) Analysis of main rocks and sediments near fluid geochemical sampling sites.

The major elements of rocks and sediments were determined by X-ray fluorescence spectrometry (XRF). The trace elements of rock and sediment samples were analyzed by element-type inductively coupled plasma mass spectrometer (ICP-MS). Refer to the relevant *Chemical Laboratory Manual* for details.

Processing steps for fluid geochemical data of hot springs, hot Wells, cold springs, and sandblasted water holes.

(1) The mantle source proportion of He and CO_2 in hot spring gas was calculated by using helium and carbon isotope in hot spring gas. The terrestrial heat flow was calculated by using helium isotopes in hot spring gases. Refer to *Isotope Mass Spectrometry Technology and Application* for details.
(2) For the calculation of hot spring water recharge elevation, thermal storage temperature, circulation depth and water rock reaction saturation index, refer to *Geothermal Geochemical Exploration* for details.
(3) The water quality assessment method is used to evaluate the water source of descending spring, water well, and tap water, with detailed reference to the corresponding *Methods and Techniques of Water Quality Analysis and The Practical Manual of Water Quality Analysis.*

2. Methods and basic requirements for geochemical investigation of soil gases in earthquake rupture zones

1) Contents of a geochemical survey of soil gas in the active fault zone.

Concentrations of He, H_2, CO_2, N_2, O_2, CH_4, Rn, and Hg in soil gas and fluxes of CO_2, CH_4, Rn, and Hg in the active fault zone.

2) The principle of laying out survey net and line.

(1) The layout should be combined with structural segmentation and fault properties, the survey lines should cross the main fault, and the observation points with fault outcrops and deformation across the fault should be selected as far as possible.
(2) Layout of lines: soil gas concentration and fluxes should be measured as far as possible at each observation site and at least 3 parallel lines should be arranged in each observation site according to the site conditions, generally 5–20 m, the distance between measuring points of each line should be: 5 m near the fault (less than or equal to 30 m from the fault), 5 m far from the fault (more than 30 m from the fault), the distance between points can be increased to 10–50 m (can be adjusted according to the topography and the type of fault).

3) Methods of geochemistry measurement of soil gases in active fault zones.

(1) The method of measuring the concentration of soil gas in the fault zone. Field sampling is drilling a hole at the measuring point (the depth is 80 cm), the sampler is placed in the hole, the hole is sealed, the air pump is used to draw the air, then the sampling measurement is carried out. The measurement requirements for each component are as follows:

① The concentration of He in the soil gas is sampled using a sampling needle or bag (which is washed several times by the soil gas), the sample should be measured in the shortest possible time, generally no more than 5 h, and measured by gas chromatography, the detection limit of the instrument should be less than 1 ppm and the error should be less than 1%;

② The concentration of H_2 in soil gas can be measured by direct air sampling or by sampling with a sampling needle or a sampling bag (to be washed several times by soil gas): The sample needs to be measured in the shortest possible time, generally do not exceed 5 h, the instrument detection limit should be less than 1 ppm, error less than 1%;

③ CO_2 concentration in soil gas is measured by direct extraction sampling, the detection limit of the instrument is less than 10 ppm, and the error is less than 1%;

④ The concentration of Rn in soil gas was measured by direct air sampling, with each measuring point measuring 15 min, one value measuring 5 min, and the third measuring value as the measuring value of each measuring point. The detection limit of the instrument should be less than 10 Bq/m^3, the error is less than 10%;

⑤ The concentration of Hg in soil gas needs to be measured by pumping air onsite. The maximum concentration measured by the instrument is taken as the measurement value of each measuring point. The detection limit of the instrument should be less than 1 ng/m^3, and the error should be less than 5%;

⑥ The concentration of CH_4 in soil gas shall be sampled with a sampling needle or bag (it shall be cleaned by soil gas many times): The sample shall be measured in the shortest possible time, generally not more than 5 h. Gas chromatography shall be used for measurement. The detection limit of the instrument shall be less than 10 ppm, and the error shall be less than 1%.

(2) The method of measuring the flux of CO_2, Hg and Rn in the fault zone: The flux of CO_2, Hg and Rn is measured by the static dark box method. The flux box is a hemispherical shell with a radius of 20 cm. The material of choice is polytetrafluoroethylene, followed by plastic or stainless steel. When measuring, select the measuring point, and start to dig the hole, 20 cm deep, 20 cm diameter, after digging, put the flux box, then use clay to seal the edge of the flux box, and finally, a loop is formed between the inlet and outlet of the instrument and the inlet and outlet of the flux box, and the temperature in the box at the time of sampling and the atmospheric temperature and pressure before and after observation are recorded. The requirements for measuring the flux of each component are as follows:

① For CO_2 flux measurement, the portable CO_2 analyzer is used to continuously pump air onsite for 30 min, and a value is measured for 10 s. At the beginning of each flux measurement, the air concentration is measured first, and the exponential fitting method is used to fit the data when calculating the flux; The CO_2 flux is calculated as follows:

$$\text{Flux}_{CO_2}\left(\text{g m}^{-2}\ \text{day}^{-1}\right) = \frac{\Delta M}{A\Delta t} = \frac{\rho_{CO_2\text{Std}} \times V_{\text{Std}}}{A} \cdot \frac{\text{d}c}{\text{d}t} = \frac{\rho_{CO_2\text{Std}} V_C P_C T_{\text{Std}}}{P_{\text{Std}} T_C A_C} \cdot \frac{\text{d}c}{\text{d}t}$$

$$= \frac{1980\ \text{g m}^{-3} \times \frac{2}{3} \times \pi \times 0.2^3 \times P_C \times 273.13}{1.01 \times 10^5 \times T_C \times \pi \times 0.2}$$

$$\times \frac{dc}{dt}\left(\frac{ppm}{s}\right) \times \frac{86{,}400\ s}{1\ day} \times \frac{10^{-6}\ m^3}{1\ ppm} \tag{6.1}$$

$$V_{Std} = \frac{V_C P_C T_{Std}}{P_{Std} T_C} \tag{6.2}$$

In Eqs. (6.1) and (6.2), c is the concentration of CO_2 in the flux box, P_C and T_C, V_C are the pressure, temperature and volume in the flux box, respectively. According to the field measurement, the standard temperature and pressure (STP) are defined as 0 °C (273.15 K) and 101.325 kPa (1 atm), 1.98 g/L (gas) (0 °C, 1 atm).

② The Rn flux is measured by pumping air continuously for 1 h with a portable Rn meter for 5 min. At the beginning of each flux measurement, the air concentration shall be measured first, and the straight-line fitting method shall be used to fit the data when calculating the flux; Radon flux is calculated as follows:

$$\begin{aligned} Flux_{Rn}\left(m\ Bq\ m^{-2}\ s^{-1}\right) &= \frac{\Delta C}{A\Delta t} = \frac{V_C P_C T_{Std}}{P_{Std} T_C A_C} \cdot \frac{dc}{dt} \\ &= \frac{\frac{2}{3} \times \pi \times 0.2^3 \times P_C \times 273.13}{1.01 \times 10^5 \times T_C \times \pi \times 0.2} \times \frac{dc}{dt}\left(\frac{Bq}{min}\right) \\ &\times \frac{1\ min}{60\ s} \times \frac{1000\ m\ Bq}{1\ Bq} \end{aligned} \tag{6.3}$$

In Eq. (6.3), P_C and, T_C and V_C are the pressure, temperature, and volume in the flux box, respectively, which are obtained through field measurement. The standard temperature and pressure (STP) are defined as 0 °C (273.15 K) and 101.325 kPa (1 atm).

③ The Hg flux is pumped continuously for 30 min using a mercury meter. The air concentration shall be measured at the beginning of each flux measurement. The straight-line fitting method is used to fit the data when calculating the flux. The flux of Hg is calculated as follows:

$$\begin{aligned} Flux_{Hg}\left(ng\ m^{-2}\ h^{-1}\right) &= \frac{\Delta M}{A\Delta t} = \frac{V_{Std}}{A_C} \cdot \frac{dc}{dt} = \frac{V_C P_C T_{Std}}{P_{Std} T_C A_C} \cdot \frac{dc}{dt} \\ &= \frac{\frac{2}{3} \times \pi \times 0.2^3 \times P_C \times 273.13}{1.01 \times 10^5 \times T_C \times \pi \times 0.2} \times \frac{dc}{dt}\left(\frac{ng}{m^3\ s}\right) \times \frac{3600\ s}{1\ h} \end{aligned} \tag{6.4}$$

In Formula (6.4), P_C and, T_C and V_C are the pressure, temperature and volume in the flux box, respectively, which are obtained through field measurement. The standard temperature and pressure (STP) are defined as 0 °C (273.15 K) and 101.325 kPa (1 atm).

6.4 Procedures

Flow chart of geochemical scientific research is shown in Fig. 6.1.

1. Preparation stage

(1) Preparation of flow measuring instruments for geochemical of soil gas in the fault zone.

Prepare a thermometer, portable gas chromatograph, hydrogen detector, helium detector, Rn detector and mercury detector.

(2) Preparation of hydrogeochemical sampling equipment.

Preparation of geochemical sampling equipment for hot spring fluid, including sampling bottle, sampling funnel, sampling pump, etc.

(3) Preparation of field geological maps.

Preparation of field geological maps, including geothermal geological maps, active fault distribution maps, and seismic activity maps.

(4) Preparation of rock geochemical sample collection.

Preparation of rock geochemical sample collection, including a geological hammer, labels, and rock sample bags.

2. Field investigation stage

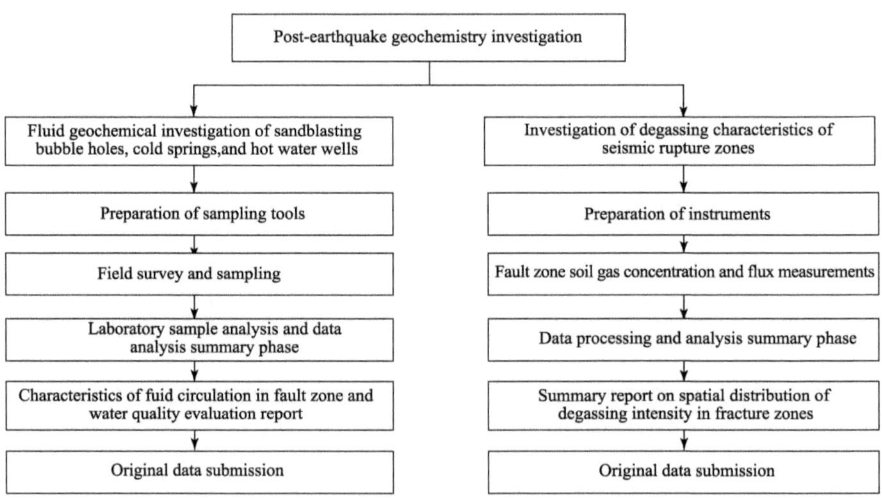

Fig. 6.1 Flow chart of geochemical scientific research

(1) Investigation and sampling of the distribution of sand-blasting water in the epicenter area.
(2) Fluid geochemical survey and sampling of hot springs, cold springs and hot water wells.
(3) Investigation and sampling of downfall springs, drinking water wells and water source areas of waterworks.
(4) Investigation and sampling of rock samples in the area with strong water and rock in the fault zone.
(5) Field geochemical measurement of soil gas in earthquake rupture zone.

3. Laboratory analysis and summary stage

(1) Laboratory analysis of sandblasting blowhole samples.
(2) Laboratory analysis of fluid geochemical samples from hot springs, cold springs, and hot water wells.
(3) Sample analysis and water quality evaluation of the water source area of the descending spring, drinking well, and waterworks.
(4) Laboratory analysis of rock samples in the area with strong water and rock in the fault zone.
(5) Analysis of geochemical spatial distribution map of soil gas in earthquake rupture zone.

6.5 Results and Data Archiving

1. Content and form of results

(1) Fluid geochemical data of sandblasting blowhole, hot spring, cold spring, and hot water well; Hydrochemical type, thermal storage temperature, water cycle depth, trace element enrichment, water–rock reaction, protolith type, regional deep mantle-derived helium distribution, mantle-derived carbon dioxide distribution, terrestrial heat flow value distribution and fluid cycle model in the seismic fracture zone.
(2) Geochemical distribution characteristics of soil gas in earthquake rupture zone, distribution map of flux release intensity of different components, environmental assessment map of release intensity of toxic gas mercury and radioactive gas Rn.
(3) Distribution map of sample analysis and water quality assessment results in the water source area of the falling spring, drinking well and waterworks.

2. Main contents of geochemical observation summary report

The report should include but not be limited to the following contents:

1) Preface.

The preface includes basic information on special topics such as work objectives, work contents, technical routes, personnel lists, and division of labor.

2) Geochemical observation.

(1) Before and after the earthquake, the fluid geochemical change characteristics of sandblasting blowhole, hot spring, cold spring and hot water well; Characteristics of deep fluid circulation in the seismic fracture zone.
(2) Geochemical characteristics of rocks in the seismic fracture zone, the degree of water–rock reaction, and the relationship between the fluid and seismic fracture
(3) The geochemical distribution characteristics of soil gas in the fault zone, the relationship between the flux release intensity distribution of different components and seismic activity.
(4) Sample analysis and water quality evaluation report of the water source area of the descending spring, drinking well and waterworks.

3) Analysis of geochemical observation results and prediction of post-seismic trend.

The fluid geochemical data before and after the earthquake are compared, and the short-term and imminent prediction opinions of seismic risk in the seismic region and its related tectonic regions are given.

4) Existing problems and deficiencies

Comprehensively summarize the work related to the geochemical observation topic, compare the work results with the expected goals, summarize the problems and shortcomings in the thematic work, and provide a reference for future scientific research work.

3. Original data submission

(1) Fluid geochemical data of sandblasting blowhole, hot spring, cold spring, and hot water well.
(2) Geochemical data of rocks in the fault zone.
(3) Geochemical data of soil gas in earthquake rupture zone.
(4) Sample analysis and water quality evaluation data of the water source area of the descending spring, drinking well, and waterworks.

Chapter 7
Source Rupture Process and Seismic Tectonic Mechanism Research

7.1 Objectives

After a strong earthquake occurred, rapid inversion of source mechanisms of the strong earthquake and aftershocks and source rupture process, and accurate location of the earthquake series can provide key information for the identification of causative source structure and mechanism, disaster rescue, seismic trend prediction and so forth. Reliable determination of causative structure is the first step to implement inversion of source rupture process from source mechanism and aftershock distribution.

7.2 Contents

Determine synthetically the nature of the causative structure and mechanism based on inversion of source mechanisms of mainshock and aftershocks, relocation of aftershock sequence, in combination with source mechanism distribution of faults near the epicenter, temporal-spatial distribution of aftershocks and on-site investigation of industrial activities.

On the basis of obtaining source mechanism of mainshock or causative source structure, obtain rapidly the preliminary result of source rupture process within several hours after the earthquake using waveform data of global or regional seismic networks, providing important references for earthquake resistance and disaster reduction. With the supplement of multi-source data, invert jointly and obtain detailed information of source rupture, providing basic data for the analysis of subsequent aftershock trend, temporal-spatial distribution of strong aftershocks and risks of surrounding main faults.

1. Source mechanism solutions of sequential earthquakes

Institute of Earthquake Forecasting, China Earthquake Administration,
A Guidebook to Earthquake Scientific Investigation,
https://doi.org/10.1007/978-981-96-0418-0_7

Source mechanism is the most important and key product besides the three seismic elements, which gives fault geometry and characteristic kinematic parameters such as rupture type and mode of source fault, fault plane parameters, source stress field and moment magnitude M_w and so forth. Mechanisms of strong aftershock sequence can provide important reference for detailed understanding of causative source structure and process and trend assessment of earthquake situation and so on.

2. Relocation of sequential earthquakes and causative structure analysis

Using travel-time data in the observation report of earthquake sequence processed from regional seismic network, relocate earthquake sequence by double-difference method, which can effectively reduce errors induced from crust structure and help to obtain better overall distribution image of earthquake sequence. Give detailed information of source structure by spatial–temporal shifting and depth profiling of location result in combination with mechanism data of mainshock and stronger aftershocks.

In case of subsequent enrichment of observation data, added by data from post-seismic closely installed emergency stations, carry out waveform correlation to obtain travel-time differences, further improve seismic relocation accuracy. Determine refined characteristics of source structure by combining geological background, mechanic solutions of mainshock, stronger aftershocks and relocation results of aftershock sequence.

3. Source rupture process

On the basis of the determination of source mechanism and causative structure, download real-time waveform data from seismic network center or the global seismic network, and preliminarily invert seismic source rupture process and give out the temporal-spatial distribution image of co-seismic static slip of strong earthquake rupture, providing reference information for determining surface co-seismic displacement positions, sizes and evaluating aftershock trends; and with the supplement of deformation observation data, directly obtain the geometric form of seismic fault, and jointly invert and obtain detailed information of fault rupture extent, providing data basis for the study of subsequent aftershock distribution and stress triggering.

7.3 Methods and Basic Requirements

1. Relocation of aftershock sequence

1) Data sources.

(1) Rapid report of earthquake sequence and catalogue results from regional networks, including seismic catalogue, observation report and waveform records.

(2) Crust velocity models. Preferably refer to the crust velocity models obtained by former researchers and without which, select the layered crust model Crust2.0.

2) Double-difference earthquake location.

(1) Select locatable earthquakes with clear phases and recorded by at least 4 stations, set neighbors and use events with minimum number of 8 phase pairs.
(2) Set maximum distance between event pair according to station distribution around the epicenter, use network catalogue as the initial locations and solve the equations by the conjugated gradient method and iterate until stable solution is obtained.

2. Mechanism solution of sequential earthquakes and causative structure mechanism

1) Method.

(1) For $M \leq 7.0$ mainshock and $M \geq 4.0$ aftershocks in the sequence, invert source mechanisms by waveform fitting method based on network waveform records with epicentral distance less than 500 km.
(2) For $M \geq 7.0$ mainshock, invert moment tensor solution by far-field seismic wave fitting method (or W-phase method), and compare it with those from other institutes such as Harvard, GCMT, GFZ etc., and analyze their reliabilities.

2) Data source and requirement.

For $M \leq 7.0$ earthquakes, download wide-band waveform data from regional networks within epicentral distance of 500 km, invert mechanism solutions by waveform fitting method (CAP, etc.) and calculate theoretical seismograms using frequency − wavenumber ($f - k$) method. For $M \geq 7.0$ earthquakes, download waveform data from the IRIS Global Network via Internet, calculate moment tensor solution by using W-phase or moment tensor inversion method from distant earthquake body waveforms.

3. Preliminary result of source rupture process

Based on layered crust model, invert and study earthquake rupture process using finite fault model and obtain rapid emergency product within several hours.

1) Data situation.

Waveforms come mainly from the wide-band data and near-field strong motion data of the China Earthquake Network (generally lacking data from Qinghai-Tibetan regions), and far-field waveform data of the Global Seismic Network.

Take far-field waveform data as an example. When an earthquake happened, download seismic data from the IRIS Data Center to study source mechanism and rupture process. Select waveform data with higher S/N rate and even azimuthal distribution (within the epicentral distance of $30° < \Delta < 90°$).

For filter processing, use suitable filtering coefficient according to data situation.

Compared to far-field waveform data, regional wide-band data of the China Earthquake Network have obvious advantages in real-time and resolution. In the case of rich observation data, carry out finite fault inversion using unspillovered regional wide-band data.

2) Waveform inversion method for source process and requirement.

Source information such as original time, place, magnitude, fault orientation, dip, and rake etc. is obtained based on the work above. Referring to source mechanism and geotectonic background of the source area, finite fault model can be constructed. The fault plane is subdivided into $M \times N$ subareas along strike and dip. After correcting corresponding time lags of subareas and by superimposing the theoretical seismogram of every subarea, theoretical seismogram of the finite fault can be calculated. Using real observation records to invert for the source parameters of every subarea, including rise time, displacements, rake and rupture propagation speed, therefore the rupture propagation process can be obtained.

4. Joint inversion of source rupture process

With the proceeding of other scientific investigations and combining on-site survey data, space geodetic data (co-seismic displacements from GNSS/InSAR, strong motion records if available), and taking advantage of multi-source data in inversion of source rupture, detailed information can be obtained based on layered crust model using non-linear inversion method (simulated annealing). Joint constraints from multi-source and independent observations are very useful for obtaining reliable source inversion result.

1) Data sources.

Data for joint inversion mainly include seismic waves and observed deformations: seismic wave data can be acquired in real-time from the China Earthquake Networks Center or the Global Seismic Network; in regions with observation conditions, it can also include mobile seismic station data from other institutions used via request. Co-seismic deformations mainly include space geodetic data and post-seismic field observation data, which are provided by deformation monitoring group and field survey group respectively.

2) Joint inversion of source process.

With the supplement of crust deformation data, causative fault geometry can be directly obtained and finite fault be constructed. Applying simulated annealing inversion, far-field displacements are calculated using layered crust model and far-field waveforms are calculated using general ray-tracing. Constructed objective function is:

$$e = \left[\frac{1}{N_W}\sum_{j}^{N_W}\left(1 - \frac{\int f_j(t)o_j(t)\mathrm{d}t}{\sqrt{\int f_j^2(t)\mathrm{d}t\cdot\int o_j^2(t)\mathrm{d}t}}\right)\right]\cdot w_1 + \left(\frac{1}{N}\sum_{i=1}^{N}\left(D_i^o - D_i^s\right)\right)\cdot w_2 + \lambda\cdot K$$

where on the right side, the first part is waveform inversion term, the second part the InSAR and GNSS co-seismic inversion term, and the third part the constraining condition (smoothing factor) used to control the continuity among neighboring subareas.

7.4 Procedures

1. Material preparation

(1) Collect relevant earthquake parameters from various institutions.
(2) Collect geological background materials of the source region, including topographic and geological maps.
(3) Collect real-time regional wide-band station records, sequence processing results, and near-field waveform records; obtain timely space geodetic observation data such as GNSS and InSAR, etc.; and collect and supplement timely data from post-seismic enhanced observation stations.

2. Source mechanisms of mainshock and sequential strong earthquakes

(1) Handle timely the automatic inversion the result from the Institute of Earthquake Forecasting, the result from artificial intelligent system of the Earthquake Science Experiment Site, and the real-time processed results from professional institutions such as GCMT, GFZ etc. Compare and evaluate the results.
(2) Download waveform data, adopt appropriate waveform fitting method according to earthquake size to obtain mechanism solutions of stronger aftershocks, and determine characteristic parameters of fault geometry and kinematics such as rupture type and mode of the causative fault, fault plane parameters, source stress field, moment magnitude and centroid depth, etc.

3. Sequence relocation

(1) Timely download and intervally update earthquake locations, phase reports from regional networks processing of earthquake sequence, adjust data formats, and analyze the data to eliminate possible erroneous data caused by artificial processing and uploading process because of seismic cluster in the initial stage.
(2) Carry out aftershock sequence relocation, correctly set parameters in every stage and adopt appropriate velocity model, and analyze the reasonability of the result.
(3) Draw plane map of earthquake sequence and analyze spatial distribution characteristics of aftershocks; make depth profile along the dominate trend of aftershocks and vertical direction, and analyze the geometrical features: strike, dip, length, fault depth and other information.

4. Determination of causative structure and mechanism

Based predominantly on the two nodal planes presented by double-couple source mechanism and referring to spatial distribution image of aftershock epicenters (after relocation), geometrical features of causative structure and mechanism can be determined, in which are included fault trend, dip, preliminary length, rupture type (strike slip, reverse, normal, etc.). Existent fault stretches around the epicenter can also provide some reference information.

5. Inversion of source rupture process

(1) Within several hours after an earthquake, obtain preliminary result source rupture process based on waveform records from regional or far-field stations.
(2) Combining multi-source data, jointly inverse source rupture process and provide result within a week after an earthquake.

7.5 Results and Data Archiving

1. Content and form

(1) Submit source mechanism solution parameters, aftershock relocation results and rupture process result.
(2) Submit maps of analyzed result of causative structure, dynamic expansion of source rupture process and co-seismic displacement field, etc.
(3) Write summary report. At the end of scientific investigation, submit summary report in the form of survey report being excellent in both pictures and composition and with relevant appendixes. Analyze and present fine characteristics of causative structure and mechanism and provide details of source rupture process.

2. Main content of the summary report

1) Preface.

The preface should include working goals, working contents and work load, investigation method and work arrangement, names of working group members and work division.

2) Data acquisition situation.

(1) Earthquake catalogues, phase data and waveform records.
(2) Other supplementary observation material.
(3) Relevant research results of other institutions.

3) Result analysis and understanding.

(1) Source mechanism of mainshock and strong aftershocks.
(2) Analyzing result of causative structure and mechanism.
(3) Inversion result of rupture process.

4) Integrated analysis.

(1) Comparative analysis of the research results from multitudinous institutions.
(2) Understanding of causative structure of strong earthquake.
(3) Details of source rupture process and slip distribution features.
(4) Trend of future earthquake sequence, possible places of stronger aftershocks and strength estimations.

5) Problems and deficiency existed.

Analyze the problems existed in the process of source inversion and relocation, and propose succeeding supplementary means; sum up comprehensively the gap between results achieved and the goals expected, and point out problems and deficiencies existed in order to provide references for later scientific investigation work.

3. Submission of original data

Submitted original data should include but not limited to the following contents:

(1) Earthquake catalogues and waveform records (far-field and regional stations).
(2) Space geodetic data—GNSS/InSAR (according to real situation).
(3) Various kinds of intermediate result maps and conclusion maps.

Flowchart of scientific investigation on source rupture process and causative structure mechanism is shown in Fig. 7.1.

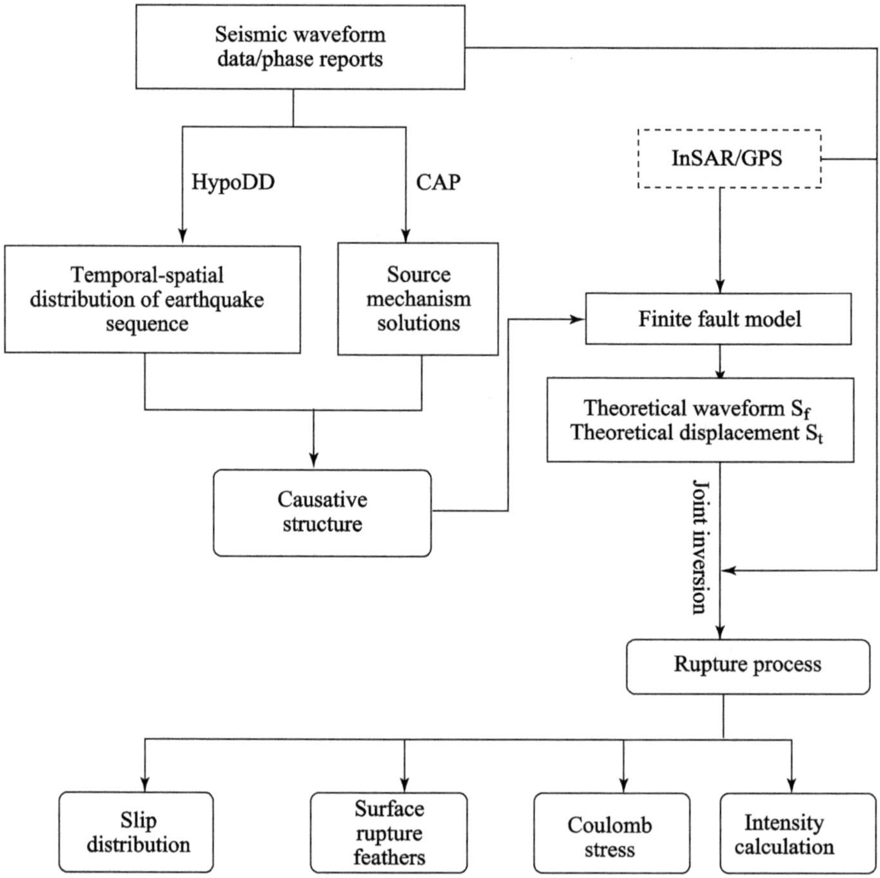

Fig. 7.1 Flowchart of scientific investigation on source rupture process and causative structure mechanism

Chapter 8
Artificial Seismic Sounding

8.1 Objectives

The high signal-to-noise ratio (SNR) observation data were obtained by artificial seismic excitation. The high signal-to-noise ratio (SNR) observation data were obtained by the artificial seismic wide-angle reflection/refraction and high resolution refraction integrated detection system, and the high resolution models of the deep and shallow crustal velocity structures and seismogenic structures in the earthquake area and its adjacent areas were constructed. Through the analysis of this model, the fine velocity structure of the crust and the relationship between deep and shallow structures in the earthquake area are analyzed, and the seismogenic structure model is established to reveal the deep tectonic background and seismogenic environment of the earthquake, which can provide background data for the post-earthquake reconstruction and the site selection of major projects.

8.2 Contents

Based on the basic understanding of the crustal structure and the distribution pattern of fault structures in the focal area and its adjacent areas, the artificial seismic large Angle reflection/refraction and high resolution refraction combined detection profiles are developed. A targeted observation system or a high-precision shallow seismic reflection detection is arranged in the shot point area and the key structural parts through which the section passes. In order to integrate deep and shallow seismic reflection and high resolution large Angle reflection/refraction detection data, study the shallow structure and deep structure differences of the focal area and its adjacent blocks, research the deep and shallow structure characteristics and mutual relationship.

Institute of Earthquake Forecasting, China Earthquake Administration,
A Guidebook to Earthquake Scientific Investigation,
https://doi.org/10.1007/978-981-96-0418-0_8

Using wide angle reflection/refraction at home and abroad for reference, and the technology of high resolution refraction detection structure and construction experience of the source area, according to the geological conditions and tectonic environment of study area, combining with the existing in the study area of deep geophysical detection results, through numerical simulation and field test, the focus of artificial seismic detection section observation system and data acquisition parameters are analyzed, The artificial seismic detection scheme suitable for the exploration research area is designed to obtain high quality deep seismic detection data.

Using advanced refracted wave velocity imaging and seismic reflection data processing methods, the data of deep and shallow seismic reflection and basement refracted wave were refined to obtain high-resolution images of subsurface structures and structures from the near surface to the Moho interface and the fine velocity structure of the basement.

Through the analysis and explanation of deep and shallow seismic reflection profile images and basement structure, considering the result of geological structure, geochemistry, geodynamics in the study area, The deep tectonic setting and shallow environment of the earthquake are analyzed, and the deep dynamic process, the shallow tectonic response and the coupling relationship of different blocks in the focal area and its adjacent areas are deduced. Then, the velocity structure of the deep and shallow parts of the crust, the basic stratification of the crust and the tectonic morphology and structural characteristics of the main interface of the crust in the study area are given. The structural styles of the main faults in the deep and shallow parts of the crust and their possible extension status to the deep are analyzed and the deep seismogenic tectonic background in the strong earthquake area is revealed.

1. Collection of previous research results in the source area

In order to better set up the artificial seismic detection and observation system, it is necessary to conduct a comprehensive review and summary of the previous geophysical research status in the earthquake area. Therefore, before the designing of the observation system, it is necessary to collect and study the artificial seismic profiles and other geophysical research results made in the earthquake area and its adjacent areas, and comprehensively analyze the crustal thickness distribution, fault distribution, tectonic characteristics, the contact coupling relationship of the block and the stress environment in the area. It provides basic data for the study of deep seismic characteristics and the boundary conditions of kinematics and dynamics, plays a very good reference role for the implementation and research of artificial seismic detection and will lay a solid foundation for further research of artificial seismic detection.

2. Observation system design and field data collection

Based on the analysis of seismic activity in the source area and its adjacent areas, the artificial seismic observation system is designed. The observation system needs to control and image the deep structure characteristics of the focal area, the distribution form of the fault and the scale of the fault. In order to improve The Times of blasting coverage in the source area and the imaging accuracy, the observation point distance and gun distance should be encrypted near the seismogenic fault.

On the basis of the perfect demonstration of the observation system and the implementation plan of the field exploration project, the collection of artificial seismic field data was organized, including the consistency experiment of seismic instruments, the layout of seismic instruments, the drilling and blasting excitation of the gun point, the recovery of seismic instruments, the arrangement and exchange of observation data, the arrangement of the field work report and other work contents.

3. Construction and interpretation of the two-dimensional crust and mantle velocity structure model in the source area

The two-dimensional fine velocity structure model of the crust and upper mantle in the focal area is constructed by processing the artificial seismic detection data collected in the field and using key processing techniques such as seismic recording assembly, seismic phase picking, ray tracing, traveltime fitting and tomographic imaging. The relationship between the crust-mantle structure model and the occurrence of strong earthquakes in the focal area is explained based on the geological tectonic environment and tectonic setting. According to the regional geological and deep structural characteristics of the study area, The deep seismic tectonic background are studied, such as the stress tectonic background of the focal area, the activity of major seismic structures and the distribution of historical earthquakes and aftershocks, the medium, geometric and dynamic conditions of the occurrence of strong earthquakes.

8.3 Methods and Basic Requirements

According to the requirements of the scientific research program, the working methods and basic requirements of artificial earthquake are determined, including the preparation of the design book and the formulation of the detection program.

1. Design and preparation of artificial seismic detection scheme

Based on the exploration task and the analysis of the collected data, the basic seismic exploration scheme was put forward, and the preliminary scheme of the location of the survey line, the observation system and the organization of the construction team was arranged. After the discussion between the experts and the members of the project team, the field exploration and technical design scheme was prepared.

The design of seismic sounding technology is the basis of seismic sounding field exploration construction. The design of the detection scheme should be determined specifically according to the detection target to be achieved by the project. The main contents and requirements of seismic large-angle reflection/refraction detection method in field construction design include:

(1) Project source.
(2) Exploration mission and research objective.

(3) Study area and survey line position.
(4) Geological structure overview of the study area.
(5) The overall design of the detection scheme.
(6) Technical requirements for field work.
(7) Field exploration construction organization and construction period.
(8) Difficulties in work and countermeasures.

2. Requirements for artificial seismic detection profile design

(1) According to the need of detecting target, the main structure of focal area is controlled effectively.
(2) The trend and position of the measurement line should be considered for the feasibility of implementation and for improving the quality of the observation record.
(3) The survey line should be perpendicular to the main structural strike and should be controlled within the effective detection zone.
(4) The survey line should pass through the major earthquake zone and the major seismic activity zone.
(5) The profile is as consistent as possible with the strike of other geophysical survey lines.
(6) The longitudinal survey line should be arranged as straight as possible, which can be properly bent or turned in the difficult section of traffic, and should be kept as straight as possible between adjacent firing points.
(7) The profile should try to avoid densely populated towns, main traffic lines and other areas with serious dynamic and human interference.
(8) In principle, the position of the shot position should be designed in a straight line with the measuring point, and the vertical deviation should not exceed the linear distance of 2–3 measuring points.

3. Design requirements of the detection profile observation system

(1) According to the design requirements of detection technology and detection targets, the design length of seismic survey line, the distribution of gun points on the survey line, the design of gun distance and the distance between observation points and other parameters should conform to the design requirements and detection targets.
(2) The length of seismic survey line should be designed reasonably according to different survey targets and different survey areas. If the above crustal structure is the main exploration target, the length of the survey line should be more than 150 km. If the crustal structure is taken as the main exploration target, the length of the survey line should be greater than 250 km.
(3) In order to continuously chase the wave sequences of the shallow, middle and deep layers of the crust, a multi-encounter and chase observation system is generally adopted.

(4) According to the technical index of the observation system design, the selected and determined shot points and observation points are marked on the topographic map or digital map along the survey line.
(5) According to the firing conditions of the blasting point, the depth of blasting materials placed in the mine and the amount of charge in a single well, the safe distance, the effective observation distance and the maximum observation distance, the explosive quantity and the combination of each gun are determined in detail.
(6) Generally, the charge of the shot point at the end of the measurement line should be appropriately increased, and the point distance of different measurement points can also be appropriately increased in the endpoint segment according to the importance and emphasis of the research.

8.4 Procedures

The workflow of the scientific research task of artificial seismic wide-angle reflection and refraction detection is shown in Fig. 8.1.

1. Indoor preparation stage of artificial seismic detection

After the earthquake, experts were quickly organized to prepare the scientific survey plan, and the Design Book of Exploration Technology was prepared according to the research objectives and technical requirements, and relevant experts were organized to demonstrate the design plan. Before the design begins, we should collect and understand the relevant information of the detection area according to the technical indicators of the assignment book. The information includes but is not limited to the following contents:

1) Natural and geographic data.

(1) The surface conditions mainly include topography, hydrogeography, distribution range and type of surface sediments in the detection area.
(2) Administrative divisions, the distribution of main traffic lines, etc.
(3) Distribution of towns and settlements, economy, culture, folk customs, etc.
(4) Climate characteristics and general situation of industrial and agricultural production in the detection area.
(5) Electronic topographic map (or topographic map) and traffic map of the detection area.
(6) Geological and seismic structure map of the detection area.

2) Seismic geological data.

(1) Overview of regional geological structure, distribution of main tectonic blocks and fault tectonic zones

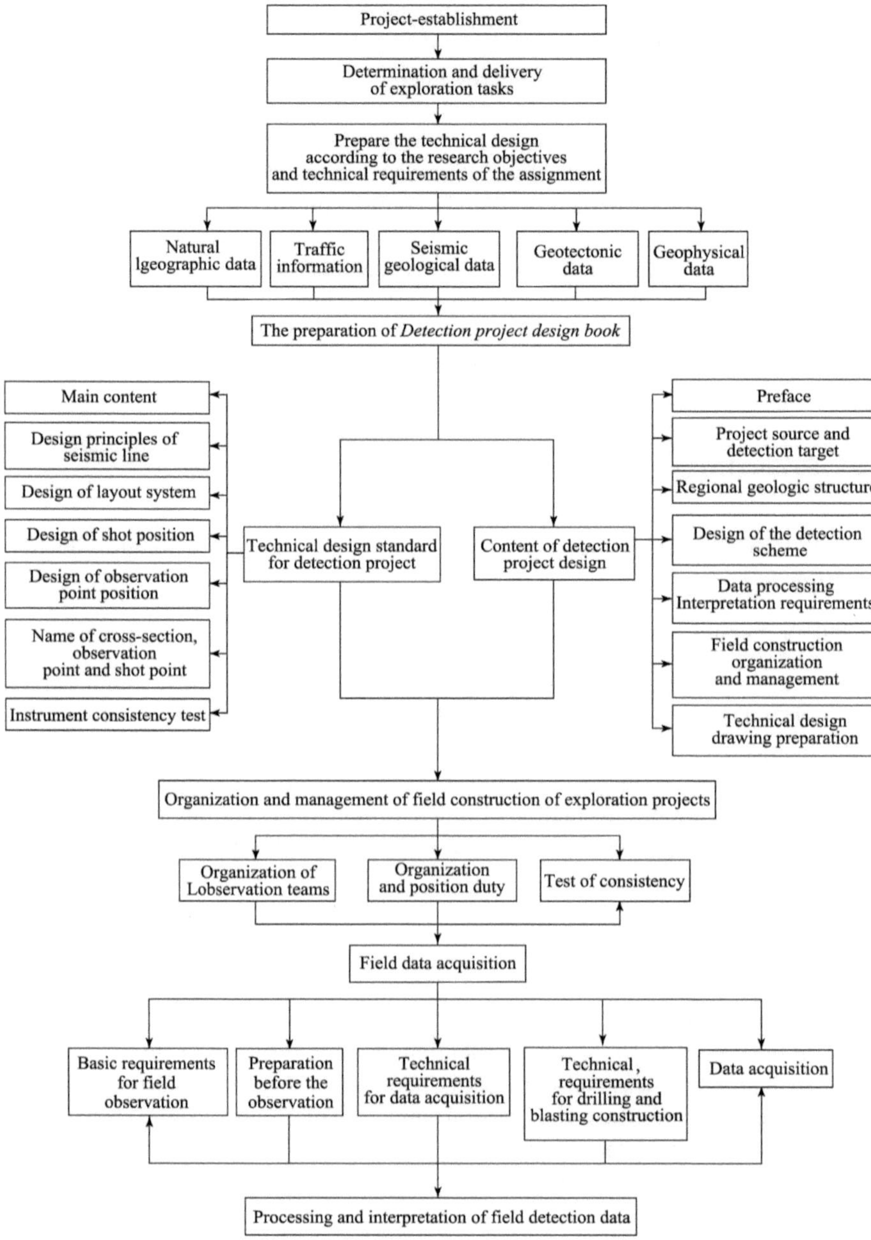

Fig. 8.1 Working flow chart of artificial seismic large-angle reflection and refraction detection for scientific research

(2) Distribution and structural characteristics of surface strata in different geological structural units.
(3) Seismic activity.

3) Geophysical data.

(1) Gravity, magnetism, electricity and other geophysical data.
(2) The main detection results of previous seismic detection profiles in the detection area and adjacent areas, includes the characteristics of seismic wave groups, the main crustal stratification and velocity data, etc.

2. Field survey stage

1) Shot position survey.

(1) The selection requirements of shot position.

Generally, the location of the shot position is relatively remote and relatively low in elevation, shallow in aquifer, and far away from the surrounding factories and villages.

(2) The excitation requirements of shot.

① If the blasting is excited in water, the depth of the water must be greater than 10 m;
② If the blasting excitation is in the borehole of the sedimentary soil layer, the blasting excitation should be in the soil layer below the submersible surface;
③ If blasting excitation is carried out in sandy and cobbly strata, large-diameter borehole, multi-node focal charge column combination binding, blasting excitation in or below aquifer can be used when the conventional drilling methods cannot drill;
④ If blasting excitation is performed in a hole in a lithospheric layer, blasting excitation is performed in or below aquifer-rich well holes;
⑤ In addition to the above conditions, the blasting excitation in shallow water or rock cave (or abandoned mine cave) can be adopted according to local conditions;
⑥ Other forms of blasting excitation should be subject to specific circumstances.

(3) Shot excitation material and well depth requirement.

① The selection of blasting excitation materials: According to the amount of single gun charge to choose different specifications of artificial seismic special source charge column. At present, the commonly used source charge column specifications (diameter/mm) are: 60, 75, 145 mm and so on. The general principle is to reduce the number of holes and the connection length of traditional Chinese medicine column in a single well as much as possible, so that the energy stimulated by blasting can be more concentrated and spread effectively.

② The choice of detonation material: Special detonators for seismic exploration with high safety factor and stable performance should be used.
③ The choice of well depth: Generally, the depth of the well is 60–80 m, the dosage of each well is about 500 kg, and the thickness of the sand layer covering the upper layer is 25–30 m.
④ In the aquifer layer, the medium density detonation velocity effect is ideal. In the rock well, the medium and high detonation velocity effect is ideal.

(4) Shot safety requirements.

① The safe distance of the blasting excitation in the well with 500–1000 kg dosage is generally safe distance: above 500 m from the ordinary residential area (general village). It is more than 200 m away from the general high-voltage lines and ordinary dykes; It is more than 150 m away from the general railway, highway and national highway.
② The safe distance of the blasting excitation in the well with the charge of 1000–3000 kg is more than 800 m from the ordinary settlement (general village), more than 500 m away from the general high-voltage line, ordinary dam. It is more than 200 m away from ordinary railways, highways and national highways.
③ The safe distance of the blasting excitation in the well with a charge of 3000–6000 kg is more than 1000 m from the ordinary settlement (general village), more than 500 m away from the general high-voltage line, ordinary dam. It is more than 350 m away from the general railway, expressway, and national highway.

2) Observation point survey.

The location selection of observation sites and the placement of seismic instruments are the key to obtaining a good seismic record. The selection should be made according to the objective conditions of different regions:

① The observation point should be as far as possible away from towns, factories, mines, trees, power transmission lines, traffic arteries and interference sources easy to cause man-made and mechanical vibration.
② Observation points located in mountainous areas and hilly areas should be arranged along the road between towns (villages) and the dirt road or service road in the mountains nearest to the survey line. As far as possible, seismic instruments should be placed in low depression, flat and hard soil layer or bedrock, and should be as far away from rivers and main roads as possible, so as to reduce the interference of river flow and mechanical power, and improve the signal to noise ratio of seismic records.
③ For some observation sites located near rivers, in low-lying areas (swamps), or in shallow areas of the water table, multi-layer plastic bags should be used when installing seismic instruments to protect them from moisture or water immersion.
④ For observation points located near plain areas and towns where river survey lines are located, attention should be paid to avoiding settlements and choosing low-lying places when installing seismic instruments. Measures such as digging deep pits, compassing the soil and burying the instruments should be taken.

⑤ To suppress wind and other natural disturbances, deep pits are dug to bury instruments when they are laid out.
⑥ The observation point should be arranged along the designed survey line as far as possible. When the interference background is large or the traffic is difficult, it can be properly offset along the designed survey line.

3. Development of the DSS detection scheme

1) The content of the DSS detection implementation plan.

(1) Layout basis of measuring line, measuring point, gun point and observation system.
(2) Location and technical requirements of gun point and observation point.
(3) Drilling and blasting mode and its technical requirements.
(4) Technical requirements for observation point selection and installation of seismic instruments.
(5) Blasting observation time arrangement and seismic signal inspection and extraction.
(6) Drilling rig type and requirements.
(7) Observation personnel and vehicle requirements.

2) Demonstration of the DSS detection scheme.

According to the field survey, the relevant experts are organized to conduct the demonstration of the field implementation exploration plan (including personnel, vehicles, instruments, construction schedule, etc.). The implementation plan is modified according to the opinions and suggestions of the experts, and organized the implementation.

3) Formation of the DSS detection scheme.

The construction plan is improved into the final implementation plan according to the opinions and suggestions of the experts. The formulation of detailed field construction scheme and related work arrangement shall be strictly in accordance with the requirements of the technical design report. Specific requirements are as follows:

(1) This section start and end coordinates and pile number and administrative place names at or above the county level along the line.
(2) The number of blasting, the equipment of the gun range, the number and model of the field receiving instrument and the distance range of the observation point.
(3) The model and number of vehicles.
(4) The start date and end time of the field construction plan.
(5) The division of labor and composition of the personnel participating in the field work.
(6) Timing of explosion.
(7) Requirements and precautions for field work.

(8) Observation point and instrument allocation for each observation group.
(9) Determining the type and number of drill rig based on the site overburden condition.
(10) HSE scheme formulation.

4. Implementation of the DSS detection scheme

1) Team organization.

In order to ensure the smooth completion of the construction task, the scientific research team organized the project team to be responsible for the field implementation of the whole project. Its organization generally include: engineering headquarters, site headquarters, drilling unit, explosion group, observation group, data group and public relations group.

2) Organization and management of field exploration and construction.

(1) Organization and staffing of field construction.
(2) Project input of the main technical equipment, budget material consumption.
(3) Production technology, economic responsibility, management mode and principles.
(4) Field construction date, process and production plan.
(5) The headquarters and each construction team shall set the location and communication information.
(6) Budget total requirements for detection works.

3) Field construction and observation.

The quality of field construction determines the success or failure of the project research, so the quality of field construction is the most important problem. The speed of the construction progress of the project and the safety of the project are directly related to the economic benefits of the project. Therefore, strict construction scheme must be formulated before field construction. Field construction production can only be carried out after the unit organizes experts to demonstrate the project implementation plan.

(1) Preparation before construction.

It mainly includes the determination of the team personnel, the type of drill rig, the number of drill rig and the construction period.

(2) Field consistency test.

① Before the field implementation of the project, the field consistency test should be conducted on all types of seismic instruments used in this time.
② The consistency test usually places all the instruments together in a circle or square, and uses dense rocks in the center of the instrument or firecrackers as a source of excitation.

③ The consistency recording profile was drawn, and the instruments whose initial signal was not obvious and the initial takeoff direction of the instrument was inconsistent were eliminated.

(3) Coordination between drilling group and explosives processing group: After the public relations and explosives procedures team makes sure, the drilling rig can start drilling. In particular, soil drilling needs to be coordinated.
(4) Excitation and observation: The site blasting plan shall be timely formulated according to the site implementation progress and the completion of the firing point, and then a notice shall be sent to the leader of the observation group through the project group.
(5) Quality timely monitoring: After the completion of the first excitation, the quality group will collect all the data recorded by the observers and conduct the recording assembly work. Instruments with poor signal–noise or instruments that have not been recorded in the recording observation section should be replaced in time to record changes in the position of observation points.
(6) After the completion of the gun point stimulation of the field project, the work of receiving point, taking number and return should be arranged in time.

4) Resettlement of a portable short-cycle seismograph. Requirements for the placement of seismic instruments:

(1) The instrument should be correctly pointed north in strict accordance with the northbound sign on the seismic instrument (It must be directed with the compass, but keep the compass away from the detector to avoid affecting the direction of the compass needle).
(2) Use the horizontal bubble that comes with the instrument to level the instrument. The instrument with the operating end should ensure that the instrument is in a level state.
(3) It is preferred to bury the instrument in the bedrock, and attention should be paid to cover the instrument to prevent interference caused by wind. In the field soil layer, the seismometer should be buried in a pit 15 cm deep to reduce noise interference, and the specific depth should be able to maintain a good GNSS signal. In all cases, the seismic instrument shall be compacted downward to ensure that the instrument base is tightly grounded.
(4) When the instrument is buried, the GNSS status should be checked to ensure that the instrument and GNSS work properly.
(5) Setting up and checking the parameters of the instrument and the observation point position, and recording the relevant parameters to make the report.
(6) After the seismic instrument is placed, the real-time waveform should be viewed to analyze and judge the environmental noise. If the field interference is too large, the interference source should be analyzed, the corresponding treatment should be made, and the better place should be selected.
(7) After the seismograph is placed, camouflage should be done to prevent the loss of the instrument. It is best to use red plastic bags or spray paint as a sign

and describe the surrounding environment of the instrument position on the notebook, to provide convenience for the recovery of the instrument after the completion of the project.

5) Drill selection.

Drilling equipment shall be selected according to the different environment, different terrain and different geological conditions of the location of shot.

6) Well hole design.

Drilling design should be reasonable to ensure the effect of blasting excitation. Considering the objective conditions of the construction area and the characteristics of blasting observation, in order to improve the blasting efficiency, Shortening the length of the medicine column makes the blasting energy more concentrated. The drilling diameter is 180–500 mm and the drilling depth is 60–80 m to ensure the excitation effect. Each downhole charge is about 500 kg. In order to ensure the blasting effect, it is necessary to fill the drilling hole to avoid well flushing during blasting. Each shot is formed by a combination of 2–4 Wells, which should be reasonably designed according to the site topography and blasting needs. In order to avoid the breakdown of borehole wall during blasting, and the energy of each well will affect each other, and the spacing between the Wells should be ensured to be 8–10 m.

Drilling must be carried out in accordance with the design requirements. If it is impossible to drill to the design depth due to complex conditions such as quicksand, pebbles and gravel layers, this can be compensated by changing the location of the shot or by increasing the number of Wells to reduce the charge per well. But only on the premise that there is no crater at the wellhead after firing and the excitation effect is ensured.

7) Loading of blasting items.

When using underground blasting excitation, due to the large amount of single cannon charge, in order to improve the blasting excitation effect, the formed seismic charge column is generally used, and the commonly used explosive detonation speed is the medium detonation speed charge column.

In order to ensure the blasting excitation effect, the following key should be paid attention to when blasting the hole:

(1) The hole must be washed repeatedly to ensure that the hole is smooth, the wall is smooth, the thickness of the sediment at the bottom of the hole is reduced, and the depth of the injection is ensured.
(2) The connection between the drug column should be tight, if the use of multi-section combination should pay attention to the binding firmly. The connection between the detonator pipeline and the explosion covered line should be tightly insulated to achieve effective waterproof effect.
(3) When the detonator is placed in the seismic charge column, strict waterproof measures must be taken, and enough waterproof materials must be used to ensure that nothing goes wrong.

(4) During the dosing process, effective measures should be taken to prevent the explosion line from being worn or hung up by the shaft wall.
(5) Ensure that the charge column is sunk to the specified drilling depth, and the well hole must be tightly backfilled with sand and stone to the wellhead after charging to ensure that the charge column does not rise.

8) Technical requirement of drilling and blasting.

(1) Construction shall be carried out according to the gun points determined by the field survey. In case of special circumstances, the gun points shall be reselected with the consent of the field technical leader.
(2) Using seismic explosive columns and detonator agreed by the field technical leader to ensure the excitation effect.
(3) The inter-well distance of combination wells is 8–10 m.
(4) The dosing depth must meet the design requirements. The drilling holes should be carried out before dosing to ensure smooth dosing. The cover layer above demolition charge is not less than 25 m.
(5) The junction between the explosive line and the detonator should be knotted. The knot should be larger than the explosive line perforation of the drug column, to prevent the joint or the detonator line from being torn off in the process of dosing.
(6) The joint of the detonator should be coated with butter as a waterproof agent. The butter should be filled at the joint to ensure the waterproof effect.
(7) After dosing, the explosive line should be placed in the top half of the well before filling, and measures should be taken to fix the wire rope at the wellhead to prevent the charge pack from sinking and pulling off the explosive line.
(8) The buried explosion line should be fixed at a distance from the well head to prevent the subsidence of the well head and the explosion line from falling into the well.
(9) In order to ensure safety and stimulation effect, the well is filled with melon seed pieces or large sand. Well filling should be carried out several times in order to better precipitate and ensure the stimulation effect.
(10) In the construction process, the relevant blasting operation rules should be strictly observed to ensure the construction safety.

5. Indoor analysis and summary stage

1) Evaluation of project quality.

The quality inspection and acceptance of artificial seismic sounding data mainly revolve around three links: technical design, field construction and data processing. Each link should be strictly in accordance with the quality standards and relevant technical requirements for acceptance and evaluation.

2) Evaluation and acceptance of original data.

Artificial seismic sounding data are evaluated according to four parts: location, drilling, explosion and observation. The evaluation is divided into three levels: excellent, qualified and waste.

3) Evaluation and acceptance of the achievement report.

The evaluation of the results report is mainly based on the technical standards and interpretation level stipulated in the technical design report. The evaluation is divided into three levels: excellent, pass and fail.

8.5 Results and Data Archiving

The work results of artificial earthquake mainly include data processing of the obtained artificial seismic observation data, model construction through a variety of data processing software, model analysis according to the obtained results, and finally give a reasonable explanation of the model.

1. Preliminary data sorting and preprocessing

1) Warehousing of the observation report.

The consistency report and geosurvey report submitted by the field observation team should be reviewed one by one to ensure that each shot and each report is correct and then submitted to the warehouse personnel for use.

2) Integration of observation data.

According to the observation report provided by the field observation group, the packing, assembly and storage of all the data collected in the field were completed completely.

3) Drawing of the record cross-section.

After filtering and correcting the original records by using the assembled data of the database, the signal-to-noise ratio of each record reaches the highest, and the cross section of the record is made according to a certain proportion of the horizontal and vertical coordinates.

2. The analysis of seismic phase or the contrast of wave group

1) The analysis of seismic phase.

(1) The seismic phase analysis should be carried out on the basis of the complete playback of the reduced record-cross section of each shot point of the whole survey line and the alignment.

(2) In seismic phase analysis, it is necessary to accurately understand the characteristics and properties of different structural units and wave groups of different seismic phases, master the change law of seismic phases, establish the spatial concept, and correctly identify the structural change characteristics of different structural units and different gun points reflected by each wave group.
(3) When comparing wave groups, different symbols or different color codes should be used to represent seismic waves of different layers and different properties (refracted waves, reflected waves, reverted waves, multiple waves, etc.). Symbols and colors used for seismic waves of the same layer should be consistent in the whole region.
(4) The following factors should be paid attention:

① The distance, range and time of different seismic waves.
② The amplitude characteristics and strong and weak energy segments of various waves.
③ The variation characteristics of apparent velocity of various waves.
④ The frequency and phase characteristics of waves.
⑤ The interference zone of several major waves.
⑦ The principle of the interchange points of traveling time.

(5) In the comparison of wave groups, the initial arrival wave and the strong energy reflected wave from the Moho near the critical point are generally compared first, then tracked to both sides, and then compared other waves layer by layer.
(6) According to the quality of the seismic phase wave group can be divided into reliable and less reliable two levels. The reliable wave group has the characteristics of prominent strength, stable waveform, obvious time interval curve, good continuity and easy to recognize. The unreliable wave group has the characteristics of unstable waveform (weak), poor continuity and not obvious features.
(7) The comparison of wave groups should be made by experienced technicians on different seismic record-cross section maps, respectively, and then determined comprehensively, or there is a person to repeat the comparison check to determine.
(8) In the process of data processing and interpretation, repeated analysis of seismic phase and wave group is needed to deepen the understanding of the reflection of seismic phase wave group on the structural characteristics of the crust and upper mantle. This observation will run through the whole process of data interpretation.

2) The collection of travel time.

When reading the effective wave travel time on the recording section, the effective wave travel time for the same level should be read carefully and accurately, and the error should not be greater than 0.05 s. The travel time of the effective wave should be as close as possible to the initial phase of the effective wave group. If it is really difficult, you can read strong amplitudes and then correct phase.

3. Data processing and analysis

(1) The velocity of seismic wave and the basic morphology of reflection interface are obtained, and the one-dimension and two-dimensional fitting calculation are carried out. At present the main used methods include WH inversion, T2-X2, PLUCH, time-term, one-dimensional traveltime fitting, finite difference inversion, ray tracing and theoretical seismogram of studying seismic wave dynamics characteristic, ZELT (forward modeling and inversion), and the theory of reflectivity and full waveform.
(2) It is necessary to have clear geological concepts that calculating and interpreting seismic data. The kinematic and dynamic characteristics are integrated to analyze and compare, and the possible underground medium structure in the study area is fully considered. Attention is paid to the range and dependency of the low speed layer (body) and high speed layer (body), so that the analysis results are more in line with the actual situation.
(3) It is a process of repeated understanding that from the seismic phase contrast to the final two-dimensional structure mapping of crust and mantle. The interpretation personnel should be familiar with the tectonic conditions and geological profiles of the study area and its adjacent areas. The researchers should make full use of other geophysical data for comparative verification and comprehensive analysis, in order to make the results accurately reflect the underground geology and structure.

4. The Content and form of seismic interpretation result

(1) The seismic record cross-section map of each gun.
(2) The positive and inversion calculation results of the one-and two-dimensional of each gun.
(3) The ray tracking, travelling-time fitting and synthetic theoretical seismic map of each gun.
(4) The two-dimensional velocity structure diagram of the crust-mantle structure.
(5) The comprehensive explanation diagram of the crust-mantle velocity structure and geological structure.

5. The main contents of the report

When all the indoor interpretation work is completed, the technical responsible personnel shall prepare the result report in time. The result report must be carefully prepared, strictly examined, and submitted in time. The summary report includes the following contents:

1) Preface.

The preface mainly includes a brief description of the task and the purpose requirements, the completion of the task and the main achievements, the construction organization and the main technical equipment.

2) The Overview of the study area.

The overview of the study area mainly includes a brief description of the location, natural conditions and economic conditions of the study area. The working conditions and their influence on the production organization and technical equipment of artificial seismic sounding need to be described. The geological structure and geophysical research of the study area and its adjacent areas are briefly described, and the past work is analyzed.

3) Working methods and techniques.

The working method and technology part mainly includes the description of the test work and its results, the observation system, the condition of the excitation reception and the working conditions of the instrument. Among them, the new technology and new method in the work are emphatically described, and the distribution and accuracy of measuring line and measuring points are described.

4) Data processing and interpretation.

(1) Review of the original data.
(2) The processing method of data.
(3) The classification and characteristics of main seismic phase.
(4) The characteristics of velocity structure and its lateral variation in the study area.
(5) The relationship between layers of media, the properties of interfaces and various structures.
(6) The contrast of the obtained results and the previous work results.

5) Conclusion and Discussion.

The conclusion and discussion part mainly describes the main conclusions of this work, the discussion of the conclusions and the suggestions for the future work.

6) The appended drawing. It mainly includes:

(1) The geological structure and line position map of the study area.
(2) The measurement line layout and observation system diagram.
(3) The seismic record cross-section map.
(4) The Ray tracing, travelling-time fitting and synthetic theoretical seismic map.
(5) The 2D velocity structure diagram of the upper mantle along the survey line.
(6) The geological profile map, the gravity and magnetic profile map.
(7) The geological and geophysical comprehensive interpretation map.
(8) Other related geophysical results map.

7) Requirement for achievement drawing.

(1) The involved geological map.
(2) Seismic record cross-section map.

(3) The 2D velocity structure diagram of the upper mantle in the study area.
(4) The location diagram of the platform array layout.
(5) The positive and inversion calculation results of 1D, 2D and 3D.
(6) The map of geological and geophysical interpretation results.

6. The submission of original data

After the completion of the project, the project overview, original data, intermediate data, achievement data and related data shall be collected and submitted within the time specified in the task statement.

1) Project overview.

The overview includes the project name and number, research objectives, scope of the research area, main geological situation of the working area, description of the observation time, main technical parameters of the source, the number of field observation sites, brief description of the observation contents, the data catalogue and the official letter of the construction unit.

2) Observed data.

Continuous waveform data recorded by each point seismometer are used in standard SEGY, DAS, or SAC format or associated standard data formats. For the non-standard data format, provide a detailed data format description need to be provided, or the batch processing software shall be provided that can transform the standard data format. The observation report generated by each measurement site shall be submitted in Word format.

3) Consistency data.

The consistent observation data of each measuring point seismometer before observation shall be provided, and the data format should be consistent with the observation data. The consistency report generated by each measuring site shall be submitted in Word format.

4) Source parameter information table.

(1) If the blasting source is used, the table includes longitude and latitude of the blasting point, elevation, zero time of blasting, well depth, number of wellhead, amount of explosive, type of explosive, conditions of the blasting site, etc.
(2) If the air gun source is used, the table includes the longitude and latitude of the air gun source, elevation, zero hour, air gun model, power, etc.
(3) If vibroseis is used, the table includes vibroseis longitude and latitude, elevation, zero hour, controllable source model, power, source output signal, etc.
(4) The source parameter information table is submitted in Word format.

5) The result table of observation site.

(1) Artificial seismic observation project information includes project name, project source, observation area, observation target, etc.
(2) The name and code of artificial seismic observation site (If the observation point is changed during the project execution, a detailed description of the change shall be given.).
(3) The longitude, latitude and elevation of the observation sites (Latitude and longitude are reserved to at least 3 decimal places, and elevation is measured in meters).
(4) The instrument model and equipment number of seismometer (the sensor type and number, the type and number of data collection).
(5) The sampling rate, the direction and polarity of the recorded channel, magnification times.
(6) The timing information of each seismometer of the observation sites.
(7) The starting and ending time of all observation sites.
(8) Instrument transfer function includes three different directions and corresponding sensitivity.
(9) The result table of observation sites shall be submitted in Word format.

6) Intermediate data and related information.

(1) The waveform data of common source. The data format is standard SEGY, USGS or Z.
(2) Picked arrive time. The data format is ASCII format.
(3) Other intermediate data and relevant information is necessary, which Other artificial earthquake science and technology projects provide according to the actual situation.

7) Outcome data and related information.

(1) The resulting data were obtained from the data processing (including 2D and 3D velocity models, velocity model data obtained after velocity analysis, data stacking, offset data, etc.).
(2) The final result data includes the resulting data of initial interpretation, the resulting data of quantitative interpretation, the resulting data of comprehensive interpretation, the imaging and modeling data, the vector diagram of interpretation profile, data quality evaluation, etc.
(3) The provision of other research outcome data, outcome maps and outcome reports obtained by the project within the observation area is encouraged.
(4) Relevant meeting notice, contract, assignment book, design book, construction scheme, implementation of project fund management, conclusion report, digital photos and drawings and other documents generated in the project declaration stage, execution and acceptance stage.

Chapter 9
Gravity Detection

9.1 Objectives

Gravity prospecting profiles or dense gravity arrays are arranged in the earthquake occurrence area to obtain Bouguer gravity anomaly and invert the 2D/3D crust in this area. The physical property distribution difference and tectonic activity of main tectonic units are analyzed to provide a reference for the study of the seismogenic tectonic environment. The data are observed through high-precision gravity data and three-dimensional coordinates of the gravity station, respectively, by relative gravity survey under the control of absolute gravity and GNSS positioning survey.

9.2 Contents

According to the actual working environment (terrain, road, residence, climate, etc.), equipment conditions in the earthquake area (gravity instruments, vehicles) and scientific research facilities, etc., geophysical data in the work area are collected, and gravity observation lines and field observation scheme need to be planned, to complete specific work such as unified data processing, output and interpretation of results.

1. observation scheme design

First, the scope of the earthquake area is preliminarily delineated. Then, collect topographic, meteorological and traffic data of the earthquake area. And secondly, design a gravity detection profile or gravity-dense array with a certain field survey. In general, the gravity profile should take the epicenter as the center and cross the main tectonic zones of important blocks in the earthquake area. Gravity dense array (composed of multiple profiles) is designed in parallel to both sides at certain intervals, taking the gravity profile passing the epicenter, to cover the entire source area as

Institute of Earthquake Forecasting, China Earthquake Administration,
A Guidebook to Earthquake Scientific Investigation,
https://doi.org/10.1007/978-981-96-0418-0_9

much as possible. The point spacing of the gravity profile or dense array is generally 0.5–2 km. Near the epicenter, the point spacing is properly densified. The spacing between dense arrays is generally 2–5 km.

2. Field work of gravity observation

Based on the above survey scheme, field gravity observation (including GNSS point measurement) is carried out. The fixed gravity measuring points of the China gravity network shall be selected as base gravity control points as far as possible to conduct a relative gravity survey of profiles or dense arrays.

The main field observations include absolute gravity measurement, relative gravity measurement, GNSS coordinate measurement, 3D laser terrain scanning (optional), etc. The specific measurement shall refer to the relevant specifications and requirements of national gravity detection.

3. Data preprocessing and gravity anomaly calculation

The free air anomaly and Bouguer gravity anomaly are obtained through correction calculation referring to the technical methods and parameters specified in the relevant national specifications for gravity exploration.

4. Comprehensive analysis and interpretation of gravity anomaly

In combination with seismic, magnetotelluric, geomagnetic and other sounding results and regional geological structures, Bouguer gravity anomaly inversion is used to calculate 2D/3D crustal density structure, including Moho undulation, give a comprehensive interpretation map (including density interface layering, density anomaly body, etc.), and analyze Bouguer gravity anomaly, crustal density structure characteristics and their relationship with earthquake preparation and occurrence.

9.3 Methods and Basic Requirements

1. Survey point/line determine

After the profile orientation or dense array area is preliminarily determined, the gravity survey points/lines shall be surveyed and selected, which can be divided into two stages: Operation on the map and field survey.

(1) Operation on the map: Preliminary setting of measuring points and lines is mainly based on traffic, topography and landform; According to the layout requirements of measuring points and regional road traffic map, the approximate location of gravity control points is preliminarily determined.
(2) Field survey: Mainly conduct field survey on survey lines and control points for operations on the map, and investigate the landform, environment, traffic, climate and other conditions in the observation area.

2. Field work of absolute gravity observation (optional)

Carry out absolute gravity measurement for all gravity control points and vertical gradient measurement at the same time to reduce the absolute gravity value to the ground joint measurement point.

1) Method.

(1) A single high-precision absolute gravimeter (FG5 or A10) is used for absolute gravity measurement.
(2) Using two high-precision relative gravimeters (BURRIS, CG5) for gravity gradient measurement.

2) Requirements.

(1) The number of effective groups of absolute gravity measurement per observation point is not less than 25×10^{-8} m/s^2.
(2) The effective number of results of single instrument for each observation point of gravity gradient measurement is not less than 5×10^{-8} m/s^2.
(3) The accuracy of absolute gravity measurement results is better than 5×10^{-8} m/s^2.
(4) The accuracy of gravity gradient measurement is better than 4×10^{-8} m/s^2.
(5) The remaining matters, in accordance with the "Crustal Movement Monitoring".

3. Field work of relative gravity observation

(1) The observation shall be conducted at a place with a wide view, flat terrain and stable foundation. Carry out relative gravity joint survey for each survey line or survey network survey point (base point network and general survey network). In each group, 1–2 sets of high-precision relative gravimeters (LCR-G, BURRIS, CG5, CG6) are used for gravity observation. The GNSS receiver is used to measure the three-dimensional coordinates (longitude and latitude, elevation) of the measuring point; At the same time, records meteorological elements (temperature and pressure). The precision of relative gravity joint measuring points should generally be better than 30×10^{-8} m/s^2, the maximum accuracy of difficult measuring points shall not exceed 50×10^{-8} m/s^2.
(2) One-way observation method can be used for field relative gravity joint survey (a closed loop shall be formed or conform to the existing control point);The length of closing or compliance time can be determined according to the performance of the instrument, generally not more than 48 h; Field observation records can be recorded in manual or electronic digital records (such as electronic records required by regulations or digital records provided by the instrument) or combination of manual and electronic records; After the completion of each survey line, the checking calculation shall be carried out in a timely manner tide correction, barometric correction, instrument height correction, zero drift correction and field checking calculation shall adopt unified field records Recording and checking procedures.

4. GNSS observation

Keep consistent with the relative gravity measuring points, and record the terrain around the measuring points (reference for near terrain correction). According to the observation conditions in the survey area, static GNSS observation or RTK observation can be adopted. The vertical accuracy of observation shall be better than 0.1 m, and the maximum accuracy of difficult measuring points shall not exceed 1.0 m. Static GNSS observation is adopted, the effective observation time is not less than 40 min, and the data efficiency is not less than 80%; The base station or network RTK shall be erected for observation, and the measurement position shall be fixed.

5. Local 3D laser terrain scanning (optional)

Based on the field survey data, the key gravity profile observation points are determined to model the local terrain using a 3D laser terrain scanner, such as areas with complex local terrain to eliminate the impact of local terrain on Bouguer gravity anomaly. The scanning range of 3D laser terrain is generally 50 m around the measuring point, and the accuracy should be better than 10 cm.

6. Data preprocessing and gravity anomaly calculation

The field gravity observation data are corrected, calculated, processed and analyzed. The calculation process is as follows:

(1) Tide, pressure, drift, and instrument height correction of field mobile gravity observation data; The absolute gravity control survey is taken as the control datum, and the observation data are preprocessed by adjustment.
(2) GNSS observation data processing and point location solution (longitude and latitude, elevation).
(3) The free air gravity anomaly of the profile or survey line is calculated by correction of normal field and elevation.
(4) The Bouguer gravity anomaly is calculated by topographic correction, and the profile gravity Bouguer anomaly curve is given.

9.4 Procedures

The basic process of gravity detection is shown in Fig. 9.1.

1. Indoor preparation stage

1) Data preparation.

(1) Collect the known parameters of this earthquake.
(2) Collect the basic parameters of main structural units and main faults in and around the seismic area.
(3) Collect geological data related to the earthquake area.

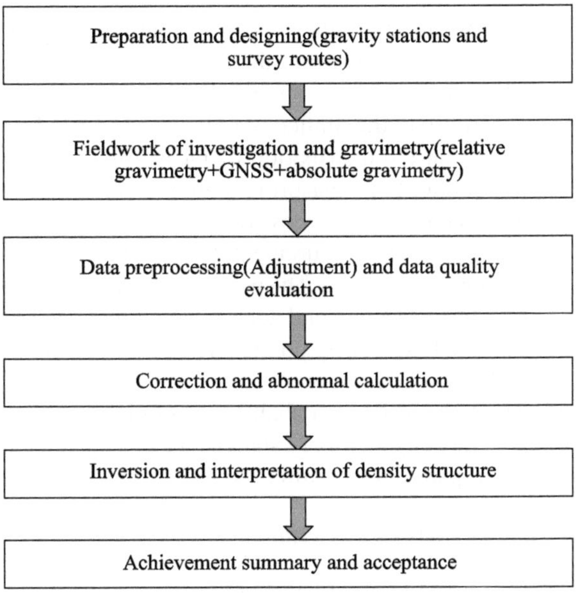

Fig. 9.1 Flowchart of gravity exploration scientific research

(4) Collect high-precision DEM data around the seismic area and survey line.
(5) Collect records of available gravity measuring points in and around the source area and recent gravity data.
(6) Work contents, priorities, and routes of field exploration shall be formulated.

2) Preliminary survey route.

Several survey lines (generally perpendicular to the structural distribution) are designed according to the distribution characteristics of the seismogenic fault and the surrounding main structures. Survey points are set according to the road distribution of the map data, and the epicenter area and the vicinity of the fault are appropriately densified; Gravity prospecting profiles or dense arrays are designed according to the structural characteristics of the epicenter area.

3) Instrument equipment and labor protection equipment.

(1) Various field observation instruments and tools, mainly including absolute gravimeter, relative gravimeter, GNSS equipment, hand-held GNSS, tripod, centering rod, barometer, thermometer, ruler, compass, laser rangefinder, 3D laser scanner, camera, wireless network equipment, etc.
(2) Preparation of various field work equipment and labor protection articles.

2. Field survey stage

According to the general requirements of the scientific research plan, this guide should be used to carry out the corresponding gravity detection.

(1) Due to the difficult working conditions, urgent time and heavy tasks of the scientific examination, the field work should be divided into several groups. The communication among the groups should be strengthened and carried out according to the unified working methods and technical ideas.
(2) Based on the distribution of measuring points in the preliminary design, the distribution of measuring points shall be adjusted according to the actual distribution of the access roads.
(3) During actual measurement, the distance between measuring points shall be as uniform as possible. The position of measuring points can be adjusted according to parking safety and terrain conditions (generally within $\pm$ 500 m of the designed measuring points). The measurement data shall be as rich as possible and recorded accurately.
(4) The field work data shall be carefully sorted and comprehensively analyzed, and the daily summary shall be made so that the field work summary report can be prepared in time after the field work is completed.

3. Indoor analysis and summary stage

(1) Gravity data adjustment processing, to obtain the absolute gravity value of each measuring point.
(2) Analysis and processing of various corrections of gravity survey data, including free air correction, height correction, grid plate correction, terrain correction, etc.
(3) The drawing and plotting of the free air anomaly map, Bouguer gravity anomaly map, normalized total gradient map of Bouguer gravity anomaly, two-dimensional crust-mantle density structure distribution map along the survey line, or three-dimensional crustal density structure distribution map of the regional crust.
(4) Comprehensive analysis of the seismogenic structure and surrounding main structures.
(5) Compile the summary report of gravity detection.

9.5 Results and Data Archiving

1. Content and form of achievements

The results of gravity exploration are mainly reflected in the distribution characteristics of the gravity field in the seismogenic region and the relationship between the gravity field and structure, the two-dimensional or three-dimensional density structure of the crust, the geometric and kinematic characteristics of seismogenic structures, the seismogenic tectonic environment and the determination of seismogenic structures. After the completion of the detection work, the final summary report and relevant annexes shall be submitted in the form of pictures and texts.

2. Main contents of the summary report

It mainly includes the actual implementation and completion of the project.

1) Data processing and exception calculation.

(1) Absolute gravity data processing method and various corrections.
(2) Relative gravity data processing method and adjustment result evaluation.
(3) GNSS data processing method and measurement accuracy evaluation.
(4) Calculation of various gravity anomalies.
(5) Inversion of 2D crust-mantle density structure in long section.
(6) Inversion of 3D crustal density structure in the seismic source area.

2) Comprehensive analysis of the seismogenic tectonic environment.

(1) The relationship between gravity anomaly characteristics and tectonic distribution.
(2) Distribution characteristics of 2D crust-mantle density structure.
(3) Characteristics of 3D crustal density structure in the hypocenter area.
(4) Comprehensive interpretation of the seismogenic tectonic environment.

3) Existing problems and deficiencies.

Comprehensively summarize the gravity exploration work, compare the work results with the expected targets, and put forward the existing problems and deficiencies, providing a reference for future scientific research work.

3. Collection and delivery of original data

The original data submitted by gravity exploration work includes but is not limited to the following:

(1) Basic geology, mapping and geographic information data involved in gravity exploration.
(2) Various original survey (mapping) data obtained in field survey, including absolute gravity data, relative gravity data, three-dimensional coordinates, near-field terrain correction data, air pressure, temperature, etc.
(3) Field paper record book or electronic record book document.
(4) Photos of measuring points.
(5) Confidential information shall be archived and managed in accordance with relevant national confidentiality management measures.

Chapter 10
Mobile Gravity Observation

10.1 Objectives

Based on the existing gravity observation facilities in the earthquake area, a certain number of absolute gravity stations and relative gravity stations are selected to establish a gravity monitoring network. Then absolute gravity observation and relative gravity observation are conducted to obtain gravity data after the earthquake. Combined with the historical gravity data in the same region, gravity variations after the earthquake are obtained, and the change of medium density and stress in the source area will be inversed and studied. The results can provide a reference to earthquake prediction and research on earthquake-generating mechanisms.

10.2 Contents

1. Observation scheme design

The mobile gravity observation area, the relative gravity stations and routes and the absolute gravity stations are determined according to the influence range of earthquakes of different magnitudes on the gravity field, the existing gravity observation facilities and the precision requirements.

Then, the working contents, the staff allocation, instruments and implementation schedule (including the observation period of the gravity network) are determined, in order to ensure the efficient completion of the scientific expedition task.

2. Field work of gravity observation

1) Absolute gravity observation.

Institute of Earthquake Forecasting, China Earthquake Administration,
A Guidebook to Earthquake Scientific Investigation,
https://doi.org/10.1007/978-981-96-0418-0_10

By using a high-precision absolute gravimeter, the gravity values of the designed absolute gravity stations are obtained, which will be used as the initial benchmarks and a control network for the relative gravity measurement.

2) Relative gravity observation.

By using high-precision relative gravimeters, the gravity differences of the designed segments are obtained. Combined with the absolute gravity observation, gravity data of the entire network are formed.

3. Data processing

After completing the field work, observation data are collected in time, including absolute gravity observation data processing results, relative gravity observation data preprocessing results and adjustment processing results. Then high-precision gravity values of all stations are obtained.

In addition, data processing should also includes reprocessing of historical data in the study area in order to obtain gravity variations of different time scales.

4. Gravity variations interpretation

The gravity variations of different time scales are systematically analyzed to obtain the characteristics of gravity change in pre-earthquake gravity variations, co-seismic gravity variations and post-earthquake gravity variations; Combined with the empirical relationship between earthquake and gravity variations characteristics, density change inversion, gravity potential energy change and other results, a comprehensively interpret the gravity variations are conducted to provide support for the study of the earthquake mechanism and the subsequent earthquake trend analysis.

10.3 Methods and Basic Requirements

1. Observation plan designing

1) Methods.

(1) The spatial range of mobile gravity observation for scientific expeditions should be determined according to the magnitude of the earthquake. It is recommended that the spatial range of gravity observation area of a $M7$ earthquake is about 450 km around the epicenter, while the spatial range of a $M8$ earthquake is about 600 km around the epicenter.
(2) The gravity observation facilities, physical geography, traffic, existing historical observation data, etc. should be collected as extensively as possible in order to provide basic data for the design of the observation plan.
(3) Considering all kinds of collected data, the relative gravity stations and survey routes should be reasonably arranged. Then, a certain number of intensified gravity stations should be added in the areas with weak monitoring capabilities.

(4) According to the principle of uniform distribution, the absolute gravity stations should be reasonably set, and additional absolute gravity stations should be added in areas with weak monitoring capabilities.
(5) According to the existing equipment and personnel, the tasks are grouped, and the personnel, equipment, and implementation plans of each group are defined.

2) Requirements.

(1) The tasks, personnel, equipment, and implementation plans of each group should be defined.
(2) A complete observation plan should be formed in order to ensure the implementation of field observation.

2. Field work of absolute gravity observation

1) Method.

(1) A single high-precision absolute gravimeter (FG5 or A10) is used for absolute gravity measurement.
(2) Using two high-precision relative gravimeters (BURRIS, CG5) for gravity gradient measurement.

2) Requirements.

(1) The number of effective groups of absolute gravity measurement per observation point is not less than 25.
(2) The effective number of results of a single instrument for each observation point of gravity gradient measurement is not less than 5.
(3) The accuracy of absolute gravity measurement results is better than 5×10^{-8} m/s^2.
(4) The accuracy of gravity gradient measurement is better than 4×10^{-8} m/s^2.
(5) The remaining matters, in accordance with the "Technical Regulations for Monitoring Crustal Movement".

3. Filedwork of relative gravity observation

1) Method.

(1) In each group, at least 2 sets of high-precision relative gravimeters (LCR-G, BURRIS, CG5, CG6) are used for gravity observation.
(2) Round-trip closed observation of A–B–C– ··· –C–B–A is used during relative gravity observation of each single survey line.

2) Requirements.

(1) The closing time of a single survey line should be less than 72 h. The closing time of a single survey line can be enlarged to 72 h under special circumstances.

(2) The self-deviation tolerance of each segment for 2 passes measuring is 28×10^{-8} m/s^2 for 2 passes, and is 35×10^{-8} m/s^2 for 3 passes measuring.
(3) The mutual deviation tolerance of each segment is 33×10^{-8} m/s^2 when 2 sets of instruments are used, and is 40×10^{-8} m/s^2 when 2 sets of instruments are used; For special difficult areas, mutual deviation tolerance can be enlarged to 1.5 times of the initial value of tolerance.
(4) For other unresolved matters, the *Technical Regulations for Monitoring Crustal Movement* should be referred to.

4. Absolute gravity measurement data processing

1) Method.

(1) According to the mathematical model of the working principle of the free-fall absolute gravimeter, the processing is carried out in accordance with the standard procedure of data reading, single-fall gravity value fitting, gravity value correction, group average calculation, and total average calculation.
(2) Correction of the solid tide, tidal load, atmospheric pressure, and polar shift are implemented in the data processing.

2) Requirements.

(1) The absolute gravity value of each observation point (instrument height and observation pier surface) and its accuracy are given.
(2) The gravity gradient measurement value between the instrument height and the pier surface of each observation point and its accuracy is given.

5. Relative gravity data processing and joint adjustment

1) Method.

(1) The solid tide correction, air pressure correction, instrument height correction, zero drift correction, etc. should be conducted on gravity observation data.
(2) Gravity difference of each segment is calculated. The self-deviation, mutual deviation, and other indicators are counted. Then the statistical table of gravity difference results is made.
(3) Combined with the absolute gravity values and the preprocessing results of relative gravity difference, an overall adjustment is carried out using the classical adjustment method.

2) Requirements.

(1) The gravity difference and accuracy of each segment should be listed.
(2) The gravity value and accuracy of each gravity station should be listed.

6. Gravity variation calculation and gravity variation image interpretation

1) Method.

(1) According to the processing results of the scientific expedition gravity observation data and the historical gravity observation data, the 0.5–3 years time scale gravity variations before the earthquake and the co-seismic gravity variations are calculated. Then, the post-earthquake gravity variations are calculated according to the observation data after the earthquake.
(2) By using the density change inversion method and gravitational potential energy calculation, expand research on gravity variations is carried out.
(3) Combined with the empirical relationship between gravity variation characteristics and earthquakes, the tectonic background of seismogenic structures, etc., the gravity variation images are comprehensively interpreted. The analysis of the seismogenic mechanism of the earthquake and subsequent earthquake prediction opinions are given from the perspective of gravity variation.

2) Requirements.

A complete research report is formed based on summarizing the gravity variation characteristics related to the development and occurrence of earthquakes.

10.4 Procedures

The workflow of mobile gravity observation is shown in Fig. 10.1.

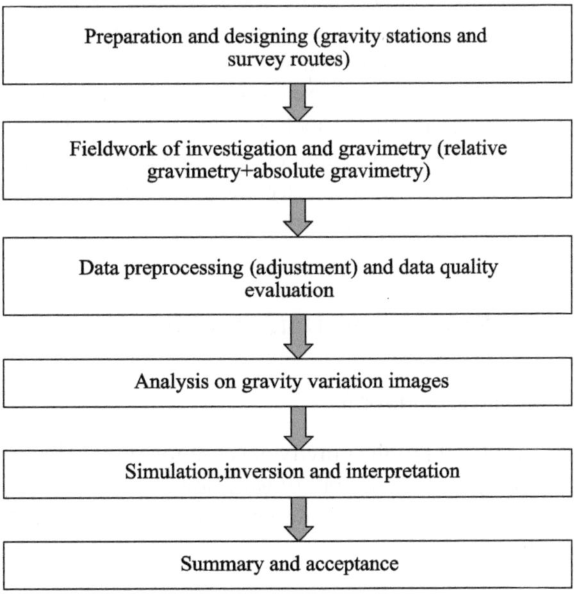

Fig. 10.1 Workflow chart of mobile gravity observation

1. The indoor preparatory phase

1) Data collecting.

Absolute gravity stations, relative gravity stations and survey routes, gravity station descriptions, and historical gravity data in the scientific expedition area are collected.

The physical geography, traffic conditions, etc. in the scientific expedition area are collected.

2) Personnel and instruments.

According to the original monitoring capabilities for the scientific expedition area, the existing available personnel and instruments, the tasks, personnel, instruments and observation plans of each task group are assigned. Then a complete Observation plan is formed.

2. The field work phase

(1) Due to the harsh working conditions, pressing time limit, and heavy tasks for the scientific expedition, communications between different groups should be strengthened in order to ensure that the field work is carried out in accordance with unified methods and technical ideas.
(2) For the purpose that the field work report can be written in time after the field work is completed, it is necessary to do some daily jobs including sorting and analysis of observation data, and making daily summaries.

3. The indoor phase of analyzing and summary

(1) After the field work is completed, the observation data should be checked in time to ensure that the quality of the observation data is qualified.
(2) The observation data should be processed in time. Gravity variation should be calculated and combined with the historical gravity data.
(3) Expand research such as inversion of density change and calculation of gravitational potential energy change should be carried out in time.
(4) Interpretation of gravity variation images should be carried out in time, then research reports are formed.

10.5 Results and Data Archiving

1. Contents and form

The results of mobile gravity observation mainly include the gravity variation images and interpretation of gravity variation characteristics. After the scientific expedition gravity observation is completed, a work report with pictures and texts should be submitted.

2. Contents of summary report

The report should include but not be limited to the following main contents:

1) Preface.

The introduction should includes work objectives, work content, work plan, work group members, and division of labor.

2) Gravity observation.

(1) Overview of the research area.
(2) Working process.
(3) Observation results.

3) Gravity data processing.

(1) Gravity data.
(2) Data processing method.
(3) Data processing results.

4) Analysis of gravity variation.

(1) The relationship between gravity variation characteristics and earthquake.

Analyzing the gravity variation images of different stages and time scales, gravity variation characteristics related to the seismogenic and occurrence process of earthquakes are summarized.

(2) Inversion and analysis of density change.

Analyzing the density change inversed from gravity variations of different periods, the material migration process related to the occurrence of earthquakes is discussed.

(3) Analysis of subsequent earthquake trend.

Based on the characteristics of gravity variations and the inversion results of density change, etc., the research and judgment opinions on the earthquake trend after the earthquake are given, which can provide a reference for the post-earthquake consultation.

(4) Analysis of seismogenesis.

The possible seismogenesis mechanism of the earthquake is analyzed base on the summary of gravity variations related to the occurrence of earthquakes.

(5) Problems and shortcomings.

A comprehensive summary of the mobile gravity observation for scientific expeditions is conducted. By a comparison of the Performance and the expected objectives, the existing problems and shortcomings are put forward as a reference to future scientific expeditions.

3. Original data archiving

The original data submitted by the mobile gravity scientific expedition gravity survey work includes but is not limited to the following:

(1) Raw data of absolute gravity observations.
(2) Outcome table of absolute gravity observation.
(3) Raw data of relative gravity observation.
(4) Outcome table of processed gravity difference of relative gravity observation.
(5) Descriptions of the changed gravity station.
(6) Adjustment processing results.
(7) The secret material should be archived and managed in accordance with the relevant rules and regulations of secrecy management of state.

Chapter 11
Magnetotelluric Sounding

11.1 Objectives

Magnetotelluric (MT) survey is carried out in the earthquake area to obtain the MT array or profile data covered the epicenter. 2D and 3D MT imaging techniques can be used to obtain deep electrical resistivity structure. The seismogenic environment and the variation of resistivity along faults at the deep depth will be revealed. The electrical resistivity structure provides deep evidence for the determination of seismogenic structure. Based on MT imaging result and the relocated seismicity, the resistivity characteristics of the earthquake focal area are revealed, and the controlling of the subsurface electrical resistivity structure combination to the main shock and the distribution of aftershock will be discussed. These works will provide a basis for discriminating the characteristics of electrical resistivity structure in seismogenic structural areas.

11.2 Contents

The first step is to design the positions of MT sites in MT array or survey line covering earthquake area. The second step is collecting MT data in the field. Qualitative analysis and inversion of the collected data are carried out subsequently. Based on the MT survey result and the existing geological and geophysical data, the characteristics of seismogenic environment and seismogenic structure should be analyzed.

1. MT dataset

Fully collect existing observation data. The positions of MT sites should be arranged according to the surface rupture of the earthquake, the location of the epicenter, the terrain conditions of the earthquake area and the traffic conditions, etc. The MT sites

Institute of Earthquake Forecasting, China Earthquake Administration,
A Guidebook to Earthquake Scientific Investigation,
https://doi.org/10.1007/978-981-96-0418-0_11

can be arranged in the form of a net array or a survey line. The data are collected subsequently in these new sites. They and the existing data compose a MT dataset.

2. Qualitative analysis of MT data

The observed MT impedance data are the direct response of the deep structure. The apparent resistivity, impedance phase, real induction vector and other data have clear physical significance. By analyzing them, the subsurface resistivity distribution in different regions and different depths can be preliminarily analyzed. The dimensional parameters, geo-electric strike and other rotational invariants can be obtained by impedance tensor decomposition. They reflect the dimensionality, dominant strike and resistivity distribution of the deep structure.

3. MT inversion

The subsurface resistivity model of the earthquake area can be established by inverted the observed MT data. The 2D MT inversion method is restricted by the assumption of 2D, and its applicability is limited in complex structural areas. 3D MT inversion method should be used for planar MT array dataset generally. The inversion of single survey-line MT data should use 3D inversion method mainly, and 2D inversion method should be auxiliary.

4. Reliability testing

The resistivity structure of the earthquake focal area is important to understand the seismogenic structure of a specific earthquake. In order to ensure the reliability of the key structure, it is necessary to test its reliability.

5. Analysis of results

Based on the resistivity model and the qualitative analysis result, combined with the existing geological, geodetic, relocated seismicity, earthquake focal mechanism and other data, the deep structural constraints on the movement and deformation characteristics in the earthquake area and the adjacent area and the seismogenic structure in the earthquake focal area should be carefully analyzed.

11.3 Methods and Basic Requirements

1. Data acquisition

1) Field measurement.

According to the tectonic characteristics of the earthquake area, the temporal and spatial distribution characteristics of aftershocks, the terrain and traffic conditions, and the seismic survey task, design the MT survey scheme with reasonable distance between survey lines and MT sites. The wide-band MT measurement systems should be used for data record. Each MT site includes 2 mutually perpendicular horizontal electric field components, 2 mutually perpendicular horizontal magnetic field components and 1 vertical magnetic field component. The horizontal component should be laid in the direction of North–South or East–West, which can be laid in a cross shape, "L" shape or "T" shape. Recording time is more than 36 h. The instruments and magnetic sensors should be calibrated before and after the recording of all MT sites, and the relative error of amplitude and phase of two adjacent calibration results should not be more than 2%. The relative error of instrument calibration, consistency comparison and checkpoint is less than 5%. Real-time field work report should be recorded for each MT site. The name of the MT sites, the coordinates, and the environment should be recorded in detail. Setup a remote reference station for continuous recording during the survey period.

2) Data processing.

The time series of each MT site should be processed by the remote reference method to suppress the noise. Choose whether to use a statistically robust algorithm according to local noise types. The high quality superposition power spectrum, apparent resistivity, impedance phase and tipper of each MT site should be obtained by careful editing. The minimum effective period of impedance is less than 0.001 Hz, and the maximum effective period is more than 300 Hz. Other requirements in data acquisition, the *Technical specification for natural source electromagnetic acquisition* (SY_ T 5820-2020).

2. Qualitative analysis of MT data

(1) Based on the distribution characteristics of apparent resistivity and impedance phase with frequency in a region, the variation of subsurface resistivity with depth in each region can be analyzed qualitatively. Based on the distribution characteristics of apparent resistivity and impedance phase at the same frequency in all sites, the variation characteristics of subsurface resistivity along the horizontal direction at different depths can be qualitatively analyzed.
(2) The real induction vector consists of the real parts of two complex numbers in the tipper. Its magnitude represents the transverse gradient in resistivity structure, and it points to the direction where the current converges. The real induction vector distribution characteristics in different regions and frequencies should be qualitatively analyzed.
(3) The geo-electric strike and dimensional parameter are important parameters to distinguish the geometric characteristics of deep linear structures. The impedance tensor decomposition technique should be utilized to eliminate the noise interference and to obtain geo-electric strike, dimensional parameter, and regional impedance of the corresponding frequency.

3. MT inversion

The MT 3D inversion method should be applied to invert the MT dataset. Different data types, such as apparent resistivity and phase, impedance tensor, impedance tensor and tipper, were selected for inversion. The results of different inversion are compared, and the differences and possible reasons are analyzed. The influence of the starting model on the inversion results should also be considered in the inversion, and the inversion results of different starting models should be compared. According to the data quality, qualitative analysis results and regional structural characteristics, the most reliable inversion results were selected as the final model. If the MT sites Arranged in a line, the inversion can be performed using the 2D inversion method and the results should be compared with the 3D inversion results. In the area with large topography fluctuation, the effect of topography should be considered in the inversion model.

4. Reliability testing

Change the resistivity values in some key areas in the resistivity model. Performed forward and inversion based on the modified model. Compare the forward response and observation data. Compare the inversion results and the prior inversion results. Determine whether the key areas in the resulting model is constrained by the inversion data.

11.4 Procedures

The scientific investigation process of MT survey is shown in Fig. 11.1.

1. Indoor preparation stage

1) Data preparation.

(1) Collect the known parameters of the earthquake.
(2) Collect MT material related to the earthquake area, including published results and existing MT data, etc.
(3) Collect the information on terrain, traffic and other conditions of the earthquake area. Design the survey coverage, the layout of MT sites, and density of the MT sites. Formulate the working route.

2) Instruments, equipment, and labor protection.

(1) Prepare necessary observation instruments, equipment and tools. It mainly includes the wide-band MT observation systems, magnetic sensors, electrodes, connecting lines, tape measures, levels, and other tools.
(2) Prepare all kinds of field work equipment and labor protection articles.

2. Field investigation stage

According to the general requirements of the scientific investigation, carry out the corresponding MT survey work in accordance with this guide.

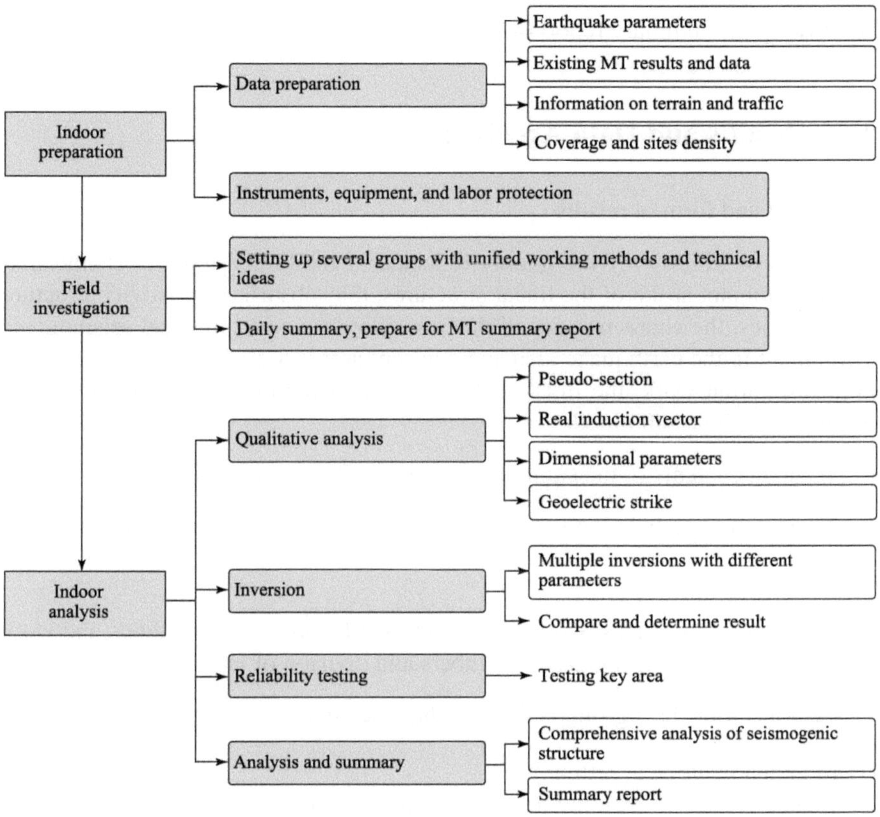

Fig. 11.1 Working routine of magnetotelluric survey

(1) Due to the difficult working conditions, urgent time and heavy tasks, field work should be divided into several groups, which should strengthen communication and carry out work in accordance with the unified working methods and technical ideas.
(2) Carefully collate the data and comprehensive analysis of the field work, and make a daily summary, so that the staff can timely prepare the field work summary report after the field work.

3. Indoor analysis and summary stage

(1) Processing and editing of observation data.
(2) Integrating the new data and the existing data to form the final dataset.
(3) Qualitative analysis of dataset, prepared the related figures.
(4) Inverting the dataset and preparing the vertical and horizontal section figures of result resistivity model.
(5) Testing the key regions of the result resistivity model.
(6) Comprehensive studying seismogenic structure.

(7) Writing the summary report of MT survey.

11.5 Results and Data Archiving

1. Content and form of results

The results of MT survey are mainly reflected in the dimensionality characteristics, the dominant strike of the linear structures, the subsurface resistivity structure characteristics, the characteristics of deep seismogenic structures and seismogenic environment in the earthquake area and adjacent areas. After the completion of the MT investigation work, the illustrated investigation report and related supporting information will be submitted.

2. The main contents of the summary report of MT investigation

The report should include but not be limited to the following main contents:

1) Preface.

It includes work objectives, work content and workload, survey methods and work deployment, list of working group members and division of labor.

2) Seismotectonic background of the earthquake area.

(1) Regional geological setting.
(2) Characteristics of regional neotectonic movement.
(3) Regional active fault distribution.
(4) Regional deep structure research and geodynamic environment.

3) MT field data acquisition.

(1) Design of the MT site positions.
(2) Field data recording situation.
(3) Data acquisition results.
(4) Preliminary and comprehensive analysis of seismogenic environment and seismogenic structure.

4) Qualitative analysis of MT dataset.

(1) Apparent resistivity and impedance phase.
(2) Real induction vector analysis.
(3) Impedance tensor decomposition.

5) Inversion and resistivity structure characteristics of earthquake area.

(1) Inversion process and comparison of different inversion results with different parameters

(2) Analysis of deep resistivity structure characteristics in earthquake area and adjacent area.

6) Comprehensive analysis of seismogenic environment and seismogenic structure.

Based on the deep resistivity structure obtained by inversion and the qualitative analysis results, the seismogenic environment and seismogenic structure of the earthquake area are analyzed by combining the existing geological, relocated seismicity, seismic focal mechanism, and geodetic data.

7) Problems and deficiencies.

Comprehensively summarize the MT work, compare the work results with the expected targets, and put forward the existing problems and shortcomings, so as to provide reference for future scientific research.

3. Data and materials submission

The data and materials submission includes but is not limited to the following:

(1) Field measurement data, including raw data and data curve figures, etc.
(2) Qualitative analysis results of MT dataset, including isofrequency map of apparent resistivity, impedance phase, real induction vector, phase tensor analysis figure, etc.
(3) Electronic file of the resistivity model in the earthquake area and its horizontal and vertical section figures.
(4) Confidential information should be filed and managed in accordance with relevant state security management measures.

Chapter 12
Deep Seismic Reflection Sounding

12.1 Objectives

With the objective of revealing the fracture tectonic characteristics of the earthquake, we set up deep seismic reflection detection lines to obtain deep seismic reflection profiles with a high signal-to-noise ratio and high resolution, and use the rich reflection seismic response on the profiles to identify the crystalline basement, Moho surface morphology, deep and shallow crustal tectonic characteristics within the detection range, especially the location of faults in the depth range of the middle and upper crust and their spatial spread, and analyze the crustal structure and deep and shallow tectonic relationships to provide a seismological basis for studying the tectonic background of the earthquake, establishing a fine crustal structure model of the earthquake area, analyzing the deep dynamics of the earthquake, and providing basic support for post-disaster reconstruction.

12.2 Contents

1. Seismic data acquisition

To fully collect and understand the research results of seismic geology and geophysics in the seismogenic area and its neighboring areas, especially the deep geophysical understanding near the detection area, and to comprehensively analyze the distribution of crustal thickness, the spreading of fractures, the geotectonic characteristics, the contact and coupling relationship of the blocks and the force environment in the area, etc., so as to provide basic data for studying the deep seismogenic structures and the boundary conditions of their kinematics and dynamics. It provides basic data for the study of deep seismogenic structures and their kinematic and dynamic boundary

Institute of Earthquake Forecasting, China Earthquake Administration,
A Guidebook to Earthquake Scientific Investigation,
https://doi.org/10.1007/978-981-96-0418-0_12

conditions and provides a reference for the development of deep seismic reflection detection.

Based on the basic understanding of geological and geophysical data in the source area and its adjacent areas, deep seismic reflection lines are laid out, controlled or explosive sources are used and targeted multiple coverage observation systems are designed in the source area and key tectonic sites through which the profiles pass to carry out deep seismic reflection detection. Using the high-resolution deep seismic reflection imaging results, we can identify the crystalline basement, Moho interface features, deep crustal tectonic features in the target area, especially the location of faults in the depth range of the middle and upper crust and their spreading patterns; analyze the fine structure of the crust, deep and shallow tectonic relationships and the tectonic background of earthquake breeding and occurrence.

According to the geological conditions and tectonic environment of the detection area, we combine the existing deep geophysical results in the detection area, optimize the observation system and data acquisition parameters through indoor numerical simulations and field tests, design a deep seismic reflection detection scheme suitable for the seismogenic area, and collect high-quality deep seismic reflection detection data.

2. Data processing and interpretation

Using advanced deep seismic reflection data processing methods, fine data processing is performed on deep seismic reflection data to obtain high-resolution images of subsurface structures and formations reflecting the near-surface to Moho interface.

The deep seismic reflection profiles obtained are analyzed and interpreted, and the results of geology, geophysics and geochemistry in the seismic area are collected to analyze the deep tectonic background and shallow fracture characteristics of the earthquake and to understand the deep dynamics and deep and shallow tectonic coupling in the seismic area and its adjacent areas. The basic stratification structure of the crust and the morphological characteristics of the main interfaces of the crust in the seismic area are given. The tectonic patterns and combinations of deep and shallow fractures in the crust are analyzed to reveal the deep tectonic background of the earthquake.

The study of the geological and tectonic environment and the geotectonic background of the source area is essential for understanding the occurrence and causes of earthquakes. Therefore, before the implementation of deep seismic reflection detection, we should have a clear understanding of the geotectonic unit in the source area, pay special attention to the collection of background data on the regional geology and deep tectonic features of the source area, and understand the stress tectonic background, the activity of the main seismic structures, and the historical earthquakes in the source area.

12.3 Methods and Basic Requirements

According to the objectives of seismic science research, measurement lines are laid across the major fracture structures where the earthquakes occurred, and the longitudinal wave reflection deep detection method with multiple coverages is used to image the fine structure of the upper mantle of the crust and the deep and shallow tectonic morphology in the source area, providing the seismological basis for the understanding of the internal structure and tectonic evolution of the earth's crust, the construction of dynamical models and the study of the seismogenic environment.

1. Deep seismic reflection line laying requirements

(1) Deep seismic reflection line should effectively cross the target tectonic system and detect the main target fault should be located in the middle of the line.
(2) The deep seismic reflection line is designed according to the requirements of the detection target and combined with the regional tectonic data, mainly in the vertical direction of the main structure.
(3) Deep seismic reflection line should be arranged in a straight line, cannot be laid out in a straight line by folding arrangement, the turning angle should not be greater than 8°, the turning point should be located in the excitation point or receiving point; cannot be laid out in a straight line and folding line, can use the curved line layout.
(4) The survey line stakes are arranged in sequence.

2. Observation system parameter requirements

1) Observation method.

The multiple coverage observation methods with internal excitation of alignment, and bilateral symmetric or asymmetric reception are used. When bilateral asymmetric reception is used, excitation is selected in the downward tilting direction of the detection target configuration.

2) Group interval.

It should be selected between 20 and 50 m, and the selection principles are as follows.

(1) Satisfying the highest non-aliasing frequency law.
(2) The lateral resolution satisfies the spatial sampling theorem.

3) Offset.

It should be determined by theoretical analysis or field tests. The minimum offset should be 0 m or one-half of the channel spacing. The maximum offset should be selected with reference to the detection of the main target layer depth, and consider the deepest target layer reflection coefficient stability within the receiving window, to meet the dynamic correction stretching distortion of not more than 12.5%, stacked acceleration analysis accuracy of not less than 6%.

4) Fold.

It should mainly improve the signal-to-noise ratio of the deep reflection signal, taking into account the resolution of the seismic record and economic efficiency and other comprehensive factors, and the fold should be not less than 48 times.

5) Instrument recording parameters.

(1) Instrument sampling interval: the sampling interval should meet the sampling theorem, in order to protect the highest effective frequency, to avoid the phenomenon of false frequency, the sampling interval should be selected as 2 ms.
(2) Instrument recording length: should be able to record the deepest target layer reflected wave data, around the tail, dynamic correction time difference, static correction time difference, and leave a certain amount of margin to meet the deepest target layer reflection imaging needs, the instrument recording length should not be less than 30 s.

6) Source parameters.

(1) Vibroseis source parameters.

① The number of combined stations and combined design graphics should be selected to meet the principle of recording reflected wave data to the deepest target layer;
② The number of vertical stacks is determined experimentally to suppress random interference, improve the reflected wave energy in the deepest target layer, and the signal-to-noise ratio and resolution of the seismic record;
③ The sweeping frequency is based on the principle that it is suitable for the response of the main stratigraphic reflections in the shell and the suppression of interfering waves, and the ratio of the starting frequency to the ending frequency is not less than 2.5 octaves;
④ The sweeping length should be conducive to improving seismic sub waves and increasing the signal-to-noise ratio of seismic records;
⑤ Output force should be no higher than 70% of the maximum static load ballast;
⑥ The sweeping method should be used to up-frequency linear sweep.

(2) Explosive source parameters.

It is advisable to use a combination of large and small doses of excitation, dose and well depth to record the target depth of the reflected wave data as a principle, determined through field tests.

① Small dose excitation should be able to record effective reflection wave data from 500 to 600 ms in the shallow part, which is conducive to excitation of broadband and high principal frequency seismic records;

② Small dose of single well excitation, the excitation depth should be selected below the top surface of the diving surface or high-speed medium and combined with the surrounding safety distance to determine, in order to suppress surface waves and acoustic waves, which is conducive to improving the signal-to-noise ratio of seismic records, the well depth should not be less than 20 m; when the diving surface is deeper, the best excitation depth should be determined by test;
③ Large dose excitation should be able to clearly record the reflected wave data in the deepest target layer, which is conducive to improving the reflected energy in the deeper part of the earth's crust;
④ Large doses of combined well excitation, combined well distance should not be less than two times the radius of the explosion, well depth should be greater than the depth of small doses of excitation well;
⑤ Detonation material should be selected with a high safety factor, and stable performance of high explosive velocity detonators for seismic exploration.

7) Numerical simulation.

When available, the design of the observation system parameters should be combined with the tectonic background data of the detection area to build a geological structure model, use ray tracing, fluctuation equations and other methods for orthogonal calculations, demonstrate the reasonableness of the selection of parameters such as channel spacing and gun check distance, analyze the characteristics of seismic wave fields in different stratigraphic structures and tectonic sites, and guide the design of the observation system parameters and feasibility analysis.

12.4 Procedures

1. Information collection phase

Quickly organize relevant experts to prepare the scientific research program after the earthquake, prepare the implementation programs for deep seismic reflection detection according to the research objectives and technical requirements, and organize relevant experts to validate the implementation programs. Before preparing the implementation programs, data related to this earthquake should be collected and understood according to the technical indexes of the mission statement, as follows:

(1) Regional geological data: geotectonic zoning, stratigraphy, lithology, tectonic features and major drilling data.
(2) Reports of previous geological research results in the detection area, papers, seismic geological maps and comprehensive research reports.
(3) Surface data: surface lithology, structure, velocity, thickness, diving surface and other hydrological data.
(4) Results of previous geophysical surveys such as gravity, magnetic, electrical, and artificial seismic.

(5) Historical earthquake data and research results in the epicenter area.
(6) Research data related to this earthquake.

2. Site investigation phase

On the basis of collecting the information of the detection area, according to the requirements of the detection task, on the satellite image or geological structure map, the preliminary design of deep seismic reflection survey line, and carry out the site investigation. The site investigation includes the following elements.

(1) Construction environments such as climatic conditions, vegetation, topography, and water distribution in the survey area.
(2) The administrative division along the survey line, the distribution of industrial and mining enterprises and settlements, the traffic conditions of roads and railroads, cultural relics and nature reserves, ethnic customs and another human geography.
(3) Ground power grid, water conservancy, military and other special facilities along the survey line, underground communication optical (electric) cable and oil, gas, water and other pipeline network facilities.
(4) Excitation and reception conditions such as near-surface lithology and dive levels along the survey line.
(5) Understand the construction technical difficulties and technical countermeasures taken to carry out deep seismic reflection detection along the survey line.
(6) Draw a sketch of the survey with line numbers and starting and ending positions to describe the survey of the line.
(7) Special areas along the survey line should be fielded to investigate other situations.
(8) The surface damage of this earthquake.

3. Preparation of implementation programs phase

According to the detection objectives and needs, on the basis of fully collecting relevant information in the detection area and carrying out on-site investigation, prepare the implementation programs, including the location of the survey line, observation system and construction team organization, etc., and determine the final implementation programs after the expert demonstration.

1) Implementation of program content.

Prepare the implementation program based on the collected data and site investigation, mainly including:

(1) Task sources and technical requirements: task sources, detection purposes, detection tasks and technical requirements.
(2) Work area tectonic background and existing work profile: work area scope, work area profile, regional tectonic background, existing work profile.

(3) Survey line layout and overview: description of the survey line location, coordinates of the start and end of the survey line, length of the survey line and the administrative names of the towns along the line, survey line control of the structure, survey area topography, traffic, climate, water system, administrative divisions and geological conditions, etc., the main difficulties and technical measures to deal with them.
(4) Seismic instruments and technical parameters: seismic instruments, recording length, sampling interval, and recording format, seismic wave excitation equipment.
(5) Observation system parameters: parameter demonstration, channel spacing, number of receiving channels, gun spacing, coverage times, offset distance, and geophone main technical indicators.
(6) Design workload and technical parameters: length of a survey line, number of physical points of excitation and reception, design well depth, excitation dosage, number of controllable seismic sources, design parameters of controllable seismic sources, etc.
(7) Data acquisition and implementation standards: measurement work and requirements, field test work, seismic wave excitation work and requirements, data acquisition work and requirements, implementation and standards followed.
(8) Data processing scheme and interpretation scheme.
(9) Output data and expected results.
(10) Safeguard measures: quality assurance measures, safety, environmental protection and health assurance measures.
(11) Task organization and management: construction organization and schedule of personnel, vehicles, equipment, etc.

2) Implementation program validation.

According to the site investigation, the organization of relevant experts for the demonstration of the detection program, and according to the experts comments and suggestions to modify and improve the organization of implementation.

4. Field exploration phase

1) Team organization.

To ensure the successful completion of the scientific research task, the organization of the undertaking unit should set up a deep seismic reflection detection command, which is responsible for the overall operation, coordination and other work of the scientific research task. Under the command department, there are field exploration command, data processing group, data analysis and interpretation group, quality assurance group and logistics assurance group, as shown in Fig. 12.1.

2) Construction management.

① Organizational structure and staffing of field construction;

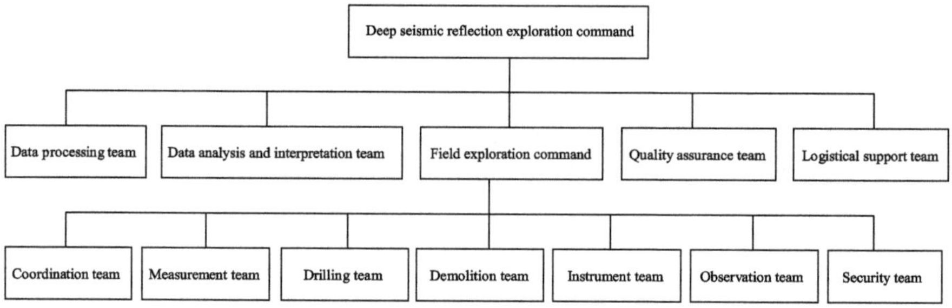

Fig. 12.1 The organizational chart of scientific research mission implementation

② The main technical equipment input to the scientific research mission, the budget material consumption;
③ Production techniques, management methods and principles;
④ Field work dates, processes and production schedules;
⑤ Command department, the construction team setup location and communication contacts;
⑥ Logistical support, and security measures.

3) Field data acquisition.

(1) Field test.

According to the geological tasks and technical requirements of data acquisition, the observation system parameters were tested for the near-surface geological and topographical conditions of the seismogenic area, with reference to the experience of reflection seismic detection in similar areas.

The field test mainly includes seismic wave excitation parameter tests, observation system parameter tests, etc.

① Field test is appropriate to use the extended arrangement method, the extended arrangement length should be greater than or equal to 1.5 times the length of the receiving arrangement, the channel spacing should not be greater than the actual working channel spacing;
② When using in-well blasting to excite, the depth of the excitation well and the amount of excitation charge should be tested;
③ When using the controlled source excitation should test the number of controlled sources, vibration force, vibration number, sweep length and sweep frequency;
④ The observation system parameter tests mainly include excitation and reception directions, number of reception channels, maximum and minimum gun detection distance, instrument recording length, and sampling interval.

(2) Seismic wave excitation.

① Seismic wave excitation requirements.

(a) Excitation should be checked before the excitation parameters, so that they meet the design requirements, after excitation, fill out the instrument shift report;
(b) Excitation should be done before the excitation point, receiving arrangements around the alert. In the case of interference sources, effective measures should be taken to exclude or attenuate the interference;
(c) The excitation point or combination of center position should be aligned with the pile, if the excitation point or combination of center position cannot be aligned with the pile, its vertical deviation should be no more than 3 times the road spacing, parallel measurement line deviation should be no more than 1/2 design gun spacing;
(d) Excitation point selection should avoid the surrounding residential factories and mines, high-voltage power grids, underground pipeline networks and other facilities to meet the safety distance required. Water facilities, railroads, fiber optic cables, gas pipelines, and other surrounding excitation point selection, to their corresponding safety distance regulations shall prevail.

② Vibroseis excitation requirements.

(a) Vibroseis should be tested for consistency before daily construction, including the synchronization of vibroseis and instruments, synchronization between vibroseis, consistency of vibroseis output signals with reference signals, and vibroseis vibration output signal waveforms;
(b) Vibroseis combination excitation should be located at the same level, and the difference in elevation within the group is not be greater than 2 m. When the difference in elevation within the group is greater than 2 m, the combination graphics should be adjusted;
(c) The phase difference between the vibroseis vibration sweep signal and the reference signal should be less than 2°;
(d) The vibrator plate of vibroseis should be kept well coupled with the ground, quality control parameters and reference technical indicators should follow the provisions of GB/T 33583-2017.

③ Explosive excitation requirements.

(a) The depth of the completed well shall meet the design well depth requirements;
(b) The borehole diameter should meet the diameter of the seismic pillar;
(c) The difference in height of the bottom of the combined wells should be less than 1 m when combined wells are used for excitation;
(d) After the seismic pillar is sunk, the pillar should be checked to see if it goes down to the bottom of the well and to prevent it from floating up;
(e) The top of the seismic pillar should be no less than 10 m from the surface, and should be located below the surface of shallow water or bedrock, and compacted with filler boreholes;
(f) Single-well excitation detonation detonators of not less than two, detonation detonators are installed in the top and middle of the pillar, using a series connection, which is conducive to the propagation of seismic waves downward;

(g) Combined well detonators should be connected in series to ensure simultaneous detonation of each well;
(h) Explosive source of excitation should fill out the gun site shift report, including the excitation point pile number, the number of wells, well depth, the number of drugs, the number of detonators and number, etc. If the excitation point pile number deviates, the direction of deviation and distance should be noted;
(i) Explosive work should comply with the relevant provisions of GB 12950-1991 and GB 6722-2014.

(3) Seismic wave reception.

① Cable seismometer reception requirements.

(a) Annual inspection of seismic instruments should be completed before construction, and monthly inspection of seismic instruments should be conducted on a regular basis;
(b) Complete daily inspection of seismic instruments before daily construction, and only put them into operation after passing;
(c) Consistency testing of the polarity of the geophone should be performed prior to construction;
(d) The cable and geophone should be checked for through-conductivity, and data acquisition can be performed after passing;
(e) For continuous empty channels or empty guns affecting the continuity of seismic records, the design of the observation system should be changed well in advance;
(f) After daily construction, the instrument operator should store and back up the instrument shift report and seismic records in a timely manner.

② Geophone placement requirements.

(a) The geophone should be well coupled with the ground to achieve the requirements of flat, stable, positive, straight and tight;
(b) Accurate location of the geophone, when it cannot be placed at the original design point, the offset along the line should be no greater than 1/10 of the road spacing, and the vertical offset should be no greater than 1/5 of the road spacing;
(c) The combination of the geophone should be conducive to suppressing the rule interference and environmental noise interference, and can be used as point combination, linear combination, or area combination; the use of linear combination or area combination should be determined by test combination direction and combination base distance, and the group height difference is not greater than the group distance;
(d) Due to surface obstacles that cannot be placed geophone, should check the empty channel pile number, and in the instrument shift report indicate the empty channel and the reason.

③ Cableless seismometer reception requirements.

(a) The charging and self-testing of the cableless seismograph should be completed before construction;
(b) After starting the cableless acquisition unit, ensure that the acquisition unit enters the data acquisition state before burying the acquisition unit;
(c) Acquisition unit embedding, should try to avoid depressions or locations prone to the formation of standing water, burial depth of not more than 30 cm;
(d) The cableless acquisition unit should be laid out according to the pile number, and the station number of the acquisition unit corresponding to the pile number should be recorded, and the acquisition unit shift report should be recorded by the observation team;
(e) After the data acquisition is completed, the data should be recovered in time. For the collection unit that cannot recover data, record it on the instrument shift report and mark its corresponding pile number;
(f) The data recovered from the cableless acquisition unit is edited and intercepted according to the designed record length to form a common gunpoint record in SEGD or SEGY format, and the common gunpoint records are arranged in order.

4) On-site quality control.

(1) Class report filling.

① Before data acquisition, the instrument operator should check the gunpoint, pile number, excitation parameters and other content; After data collection is completed, the operator should fill out the instrument shift report in a timely manner.
② The instrument shift report should be filled out completely, neatly and accurately, including the survey line number, detection direction, file number, road spacing, gunpoint stakes, alignment stakes, instrument recording parameters, excitation type and excitation parameters, etc.
③ Instrument shift report file numbers should be recorded in order, with notes on the shift report for spent guns and reacquired seismic records.
④ The observation system change, empty channel, empty gun, interference, etc. in data acquisition should be stated on the instrument shift report.

(2) Quality control.

① Before the daily construction, the instrument system, vibration source system, and power supply system should be tested, normally before the construction.
② Before excitation, the instrument operator should check the excitation point pile number, alignment pile number, excitation parameters, instrument recording parameters, etc., and be accurate before construction.
③ Irregular working channels and empty channels appearing on the seismic records should be verified and processed in a timely manner.

④ During data acquisition, the instrument operator should monitor the background noise, observe whether the auxiliary channel is normal, whether the excitation position is accurate, whether the change of the initial arrival time of adjacent channels is normal, whether the records of each seismic channel are normal and whether there are changes in the quality of the records, etc. The problems found should be dealt with and eliminated in a timely manner, and the seismic records should be reacquired if they are not qualified or abnormal.
⑤ After the daily construction is completed, the cable seismograph should backup and copy or transmit the raw data and instrument shift report to the quality control personnel for on-site processing.
⑥ After the daily construction is completed, the cableless seismograph should upload the data of the acquisition station deployment to the host computer in time, and check and inspect the data.

5) On-site data pre-processing.

(1) Basic requirements.

① Field construction period should be equipped with a field processing system to monitor the quality of field construction, test the effectiveness of the test, guide field construction operations, and provide timely data on preliminary results;
② The field processing system should be equipped with the basic modules of the regular seismic data processing program, including pre-processing, spectrum analysis, channel editing, initial-to-exit, amplitude compensation, filtering, channel extraction, static correction, velocity analysis, dynamic correction, horizontal superposition, post-stack modification and profile display, etc.;
③ The data should be collected on the same day and the initial stacking processing should be completed on the same day, so as to guide the field construction on the next day.

(2) Data processing requirements.

① The observation system is correctly defined;
② Preliminary wave removal appropriate;
③ The processing parameters of the main processing links (denoising, amplitude compensation, filtering parameters, static correction, etc.) are determined by testing;
④ Reasonable choice of stacking speed;
⑤ Clear, energy-balanced, high signal-to-noise ratio and resolution in the initial stack profile.

6) Quality evaluation of raw seismic records.

According to the quality of the raw seismic records, the quality of the records should be evaluated according to “excellent”, “qualified” and “unqualified”. The qualification rate of the raw seismic records should be no less than 90%.

(1) Excellent.

① The rate of empty channels caused by bad channels, abnormal channels or non-obstacle causes on seismic records is not greater than 1%;
② High signal-to-noise ratio of seismic recordings, significant first-to-wave, and effective suppression of random and surface wave interference;
③ Seismic records with high resolution, reliable identification of shallow reflection wave groups, and clear in-shell reflection wave groups;
④ Deep Moho and upper mantle reflection data can be clearly identified;
⑤ The initial stacking time profile reflects a clear wave group with balanced energy, a high signal-to-noise ratio and high resolution.

(2) Qualified.

① Empty channel rate greater than 1% due to bad, abnormal or non-obstructive channels on seismic records;
② Average signal-to-noise ratio and strong interference waves, which can identify the reflected wave group of the target layer after preliminary processing;
③ The initial stacking time profile reflected wave group can be resolved, with average energy and average signal-to-noise ratio.

(3) Unqualified.

① Raw seismic records with low signal-to-noise ratio and insignificant reflected wave data that are still unusable after initial processing;
② Seismic records with incorrect parameter selection and unusable data processing;
③ The initial stacking time profile reflected wave set is blurred and the signal-to-noise ratio is low.

5. Data processing phase

1) Data processing content.

The processing flow and parameters shall be determined by experiments according to the geological and tectonic characteristics of the detection area, the task requirements and the characteristics of the raw seismic records, and quality control shall be performed during the data processing. Data processing should include data format conversion, observation system definition, polarity regulations, amplitude compensation, trace shot editing, static correction, prestack noise elimination, deconvolution, velocity analysis, horizontal stacking, migration, modified processing, filtering and gain display in 13 aspects.

2) Data processing requirements.

(1) Data format conversion.

Decompile the raw seismic data format into the data format used by the seismic data processing system.

(2) Observation system definition.

① Define the excitation and reception points should match the actual field construction records;
② Check the correctness of the definition of the observation system using relevant technical means;
③ For curved survey lines, the location of the inflection points and CMP surface elements, the number of strips and the longitudinal and transverse span should be reasonably selected to make full use of the seismic data, so that the number of CMP surface elements covered and the gun spacing are as uniform as possible.

(3) Polarity regulations.

① If the initial to downward jump (negative value) is recorded, it is called normal polarity; conversely, it is called reverse polarity. In data processing, normal polarity is usually used if not specifically stated;
② For the measurement lines that do not clearly provide polarity data, the polarity should be identified by zooming in to show a small number of single gun records. For reverse polarity records, reverse polarity processing should be performed.

(4) Amplitude compensation.

The amplitude compensation method combining spherical divergence compensation and surface consistency compensation should be used. After the seismic record is amplitude compensated, the energy in the shallow, middle and deep layers should be basically balanced.

(5) Trace shot editing.

Reject bad and abnormal channels manually or by using software, remove wild values by time windows, and reject waste cannons. When using automatic channel editing technique, you should compare the seismic records before and after editing without damaging the valid messages.

(6) Static correction.

① Based on the planar position and elevation of the excitation and reception points, low drop velocity zone data, diving surface data, etc., choose the appropriate method to carry out the calculation of the static correction volume in the field, as appropriate, and compare it with the raw data and superimposed profiles without applying the static correction volume, and the effect should not become worse;
② For data with drastic surface changes or large lateral changes in the low descent zone, when the field static correction effect does not meet the requirements, the appropriate refraction static correction method should be selected, it is recommended to use the laminar static correction method, static correction calculation, and comparison with the data and superimposed profiles without applying the refraction static correction amount, the effect cannot become worse;

③ The residual static correction should be carried out iteratively with the velocity analysis, and the calculation window of the residual static correction should be selected on the seismic layer with better reflection quality, and the final maximum residual static correction should be no more than 1 sampling interval. After each residual static correction, the residual static correction effect profile should be displayed, and the quality of the residual static correction profile should be better than the profile before the residual static correction. The number of residual static correction and velocity analysis iterations should be no less than 2 times.

(7) Prestack noise elimination.

① The use of filtering over other effective means, according to the characteristics of interference waves, the use of multi-domain multi-method denoising technology, suppression of surface waves, linear interference waves, sound waves and other noise interference;
② Show some of the single gun records and the first stack of profiles to compare and check the denoising effect;
③ After noise attenuation, the signal-to-noise ratio of the seismic data should be improved, the wave group characteristics are clear, and there is no obvious effective signal in the removed noise data.

(8) Deconvolution.

The method and parameter selection should be tested to improve the resolution of the seismic record without reducing the signal-to-noise ratio of the deep effective reflection. It is recommended to use the surface consistent deconvolution method and compare it with the superimposed profiles without applying anti-folding treatment, with no reduction in the effect.

(9) Velocity analysis.

① The interval of velocity analysis points cannot be greater than 50 CMP, and the interval of velocity analysis points should be reasonably adjusted according to the geological structure, and can be appropriately encrypted when the geological structure changes drastically and requires encrypted interpretation.
② The number of CMP summations used for velocity analysis points is reasonable and should include as many tracts as possible with different shot-geophone distances. Correct parameters for dynamic correction of stretch excision and protection of shallow data.
③ The velocity analysis should refer to the results of geological formations and velocity scans. The velocity scan range should be larger than the velocity range present in the actual data.
④ The velocity pickup is reliable and the spatial variation is reasonable. Attention should be paid to distinguish the performance characteristics of multiple waves on the velocity spectrum for attenuation of multiple waves.

(10) Horizontal stacking.

① Select appropriate normal moveout and stacking methods;
② The cutting parameters are correct, and the stacking velocity field is reasonable;
③ The quality of the final stacked profile should be better than that of the intermediate stacked profile.

(11) Migration.

① Do the migration method and velocity field test, and determine migration method and migration velocity field according to the experimental results;
② The migrated result profile has reasonable event homing, clear breakpoints, and no spatial aliasing and arcing affecting seismic interpretation;
③ If conditions permit, pre stack time migration or depth migration method can be used for migration and compared with post-stack migration effect.

(12) Modified processing.

① The data after superposition and offset can be moderately processed to improve the signal-to-noise ratio;
② The profile after the improved signal-to-noise ratio processing does not blur the breakpoints, does not produce construction artifacts, and has no false frequency phenomenon.

(13) Filtering and gain display.

① The profiles displayed after filtering and gaining, with various tectonic phenomena clear and easy to interpret;
② Longitudinal and transverse display ratio determined by test, should make the main structure output display clear.

3) Processing quality inspection.

(1) Quality inspection basis.

The processing quality inspection shall be based on the technical indicators, technical requirements and relevant technical standards specified in the scientific research assignment.

(2) Quality inspection contents.

① Check the quality control drawings and intermediate results of each step of the processing flow to ensure that the selected processing flow is correct and the processing parameters are reasonable;
② Compare the original processing section with the final processing section to see whether the effect is improved; Check the final treatment profile to see if it meets the geological objectives and contract requirements.

4) Processing result data and format.

(1) The data processing result data contains the following contents.

① Data collection test typical single gun records;
② Typical single-shot records before and after data processing;
③ The average velocity data body of the deep seismic reflection profile;
④ Superimposed time profile data body;
⑤ Post-stack offset profile data body;
⑥ Processing summary report.

(2) Data processing result format.

① The superimposed time profile data body is in SEGY format;
② The data body of the post-stack offset profile is in SEGY format;
③ The final result data should contain the coordinates of the CDP points in the header, where the *X* coordinate should be located in 73 bytes of the header and the *Y* coordinate in 77 bytes of the header.

6. Data interpretation phase

Data analysis and interpretation should be carried out by a combination of superimposed and offset profiles. The reflected wave group characteristics on the profile should be combined with geological data and other geophysical data for comparative analysis of reflected waves to determine the relationship between the seismic wave group and the geological layers or formations in the shell, the distribution range and variation pattern.

1) Seismic reflection layer naming.

(1) Marking of seismic geological horizon of systems and series.

The seismic reflection layer of the system or series is defined as the reflection at the bottom interface of the corresponding geological section. The seismic geological horizons code of the seismic reflection layer is identified as T_A, where A indicates the horizon code of the system or series.

(2) Marking of seismic geological layer on the basement surface. The seismic geological layer code of the bottom is T_g.
(3) Mark of seismic geological horizon of reflection layer of special geological interface.

The seismic geological layer code of the reflection layer of the special geological interface can be determined according to the above principles to distinguish the geological meaning represented by the reflection layer.

2) Seismic reflection layer quality evaluation.

The evaluation is carried out in layers according to the needs of seismic interpretation, and the evaluation is divided into three levels:

Level I: Seismic reflection wave groups are clear and can be reliably tracked by comparison;
Level II: Seismic reflection wave groups can be identified for comparison;
Level III: Seismic reflection wave groups are difficult to trace for comparison.

3) Analysis and interpretation of seismic data.

(1) Comparative interpretation of reflected wave group.

① Comparative tracing of reflected wave groups based on horizontal superposition profiles and offset profiles compared with each other;
② In the horizontal superposition profile, the bypass waves and cross-section waves should be identified;
③ Identify the main reflection layers within the upper crust;
④ Applying wave dynamics and kinematic characteristics, taking into account the comparative interpretation of shallow, middle and deep layers;
⑤ Identify the main tectonic phenomena with reference to the background of tectonic evolution in the work area;
⑥ Identify the main structural dividing lines such as crustal structural stratification and Moho surface.

(2) Interpretation of faults and structures.

① On the horizontal stacking and offset profiles, identify positive or reverse faults based on the fault identification markers of the reflector layers;
② On the profile, to identify the reliable location of the upper extension of the fault and the lower extension layer and depth;
③ Identify the relationship between deep and shallow structures in the profile with reference to the background of tectonic evolution in the work area;
④ Comprehensive analysis recognizes the nature, level, spatial combination relationship of faults, and the controlling effect on tectonics. It can be divided into three levels according to the distribution and control of faults in the plane, namely, primary faults (faults controlling the boundaries of basins, depressions or sags), secondary faults (faults controlling the development and formation of secondary tectonic zones),tertiary faults (faults controlling local fault blocks, trap, high points and sporadic faults).

4) Comprehensive explanation.

The comprehensive interpretation should be carried out in conjunction with regional geological and other geophysical data, mainly including the following.

① Fine stratification structure and characteristics of the crust and upper mantle;
② The nature of faults and tectonic spreading characteristics;
③ Characteristics of deep and shallow tectonic relationship;
④ Tectonic evolution and seismogenic background understanding.

5) Confirmation of the reasonableness of geological interpretation of reflection seismic data.

① The seismic geological horizon is correctly calibrated;
② Reasonable interpretation of the nature of faults;
③ Reasonable interpretation of the control role of faults on tectonics;
④ Reasonable interpretation of fault drop, intersection and cutting relationship;
⑤ The interpretation of crustal structure stratification and Moho surface is reasonable.

12.5 Results and Data Archiving

1. Results output

1) Data results.

(1) Raw seismic data in the format of SEGY.
(2) Typical records and profiles of intermediate processing in SEGY format.
(3) Stacked time profile data body in SEGY format.
(4) Post-stack offset profile data body in SEGY format.

2) Material results.

(1) Implementation program of scientific expedition tasks (see the content of Implementation Programming).
(2) Original field records: instrument shift report and artillery field shift report.
(3) Field survey results: excitation point coordinates, receiving point coordinates and elevation data.
(4) Work photos, video materials, approval letters, acceptance materials, etc.

3) Report results.

(1) The structure of the report of deep seismic reflection exploration should include: preface; job content; Attached drawings and schedules; reference; Database detection report. The foreword should reflect the source of the task, the unit undertaking the task, the person in charge, the main participants, the starting and ending time and the completion time of the task.
(2) The contents of the scientific research report on deep seismic reflection exploration should include: Objective tasks; Technical requirements; Detection method; Data acquisition; Data processing; Data analysis and interpretation; Conclusion and understanding of scientific research. Among them, data acquisition should include survey line location and geodetic work, technical difficulties and countermeasures, use of instruments and equipment and parameters, field test and analysis, observation system parameters, completed workload and

quality evaluation; The data processing shall include the processing methods and processes adopted, the selection of processing parameters, and the comparative analysis of processing effects.

4) Results map and results table.

(1) Results map.

① Location map of deep seismic reflection line;
② Velocity structure profile;
③ Stack time profile;
④ Depth interpretation profile;
⑤ Comprehensive interpretation result map.

(2) Results table.

① Workload statistics;
② Statistical table of coordinates of excitation and reception physical points;
③ Elevation statistics table.

2. Data archiving

1) Ways of archived data.

After the end of the seismic research mission, the data should be archived in time. The data archiving should include paper media and digital storage media archiving.

2) Classification of archived data.

The archived data shall conform to the data entry format requirements and be classified as follows:

(1) Data type: raw single shot seismic data, original single shot combined seismic data, reflection stack time profile data.
(2) Report forms: instrument shift report, artillery field shift report, observation group shift report, physical point measurement data, survey line terrain and surface feature description.
(3) Graphics: survey line location map, velocity structure profile, stacked time profile, depth interpretation profile, comprehensive interpretation profile and typical image data during task execution.
(4) Documents: implementation programs, survey summary report, field completion report, technical report, database inspection report, relevant letters during task execution, acceptance comments on field data collection and task acceptance.

Chapter 13
Mobile Seismologic Array Observation

13.1 Objectives

The objectives of this task are to improve the capability of earthquake detection and source parameter inversion, as well as the precision of earthquake location and the resolution of subsurface imaging, by rapidly deploying seismic arrays in and around the focal area soon after the main shock. The improvements will result in more complete and accurate 4D earthquake activities, fine 3D subsurface structure and stress field, which will shed light on the deep seismogenic environment and provide the background for analyzing the earthquake trends, understanding fault anatomy and looking backward at the mechanisms behind earthquake nucleation.

13.2 Contents

To achieve the objectives, we need to design and set up a temporary seismic network on the basis of the distribution of existing seismic stations. This will generally include two kinds of deployments. To enhance the aftershock monitoring, several seismic stations will be installed to update the aftershock zone station coverage and will be run for a longer period to follow the aftershock sequence. At the same time, to increase the spatial sampling and improve the subsurface imaging resolution, a dense seismic array with a relatively short running period can be also carried out. Finally, on the basis of the above observations, comprehensive seismological studies are conducted.

1. Aftershock monitoring enhancement network

In accordance with Chap. 5 of the "Regulations on the management of earthquake site work" requirements for temporary seismic monitoring, the observation plans should be made quickly and the field work should be carried out as soon as possible.

Institute of Earthquake Forecasting, China Earthquake Administration,
A Guidebook to Earthquake Scientific Investigation,
https://doi.org/10.1007/978-981-96-0418-0_13

The continuous waveforms of the new stations must be transmitted in real-time to the data centers of the local Provincial Earthquake Agency and the China Earthquake Network Center and processed jointly with the permanent seismic network. For a more comprehensive capture of aftershock sequences, it is necessary to increase the station number and prolong the running period of the temporary network.

2. Dense seismic array

The dense seismic array aims to reconstruct high-precision temporal-spatial seismicity and to generate a high-resolution image of the seismogenic environment and fault structure in the focal region. The array geometry will be planned according to the strike of the fault and the distribution of earthquakes, which will be densely distributed with short intervals around the seismogenic fault in conjunction with the current seismic stations. Intensive aftershock monitoring in the epicenter zone will provide the observational basis for a comprehensive examination of deep structures.

3. Earthquake sequence study

Using the catalog supplied by the network center and missed earthquakes discovered by rescanning continuous waveforms of temporary and permanent seismic stations with contemporary detection techniques, a more comprehensive earthquake catalog will be compiled.

On the basis of the updated 1D or 3D velocity structure, the new catalog will be relocated in absolute or relative fashions. Using a dense seismic network and integrating inversion with waveform, first arrival polarity and amplitude information, the focal mechanism solutions or stress drops of earthquakes can be significantly enhanced. Consequently, the temporal-spatial patterns of seismic activity and stress progression of the earthquake sequence will be retrospectively evaluated, and the deep shape of the seismogenic fault will be defined.

4. Deep seismogenic environment research

Using the traveltimes, amplitudes, or waveforms of seismic events or surface wave dispersion curves derived from background noise, the 3D medium velocity and attenuation structures of the fault zone and the focal area may be concurrently or separately inverted. The source zone stress field can also be derived from the earthquake sequence's focal mechanism solutions. By combining the 3D structures, the stress field, and the relocated earthquake distributions, we can analyze the relationships between the earthquakes, the fault structure, and the stress field and investigate the seismogenic conditions and nucleation mechanism in greater detail.

13.3 Methods and Basic Requirements

1. Aftershock monitoring enhancement network

The network is intended to enhance the earthquake monitoring capability of the existing permanent seismic network in and around the epicenter, particularly for enhancing the detection of aftershocks and their precise location. Once the temporary seismic stations are installed at the designated locations, their recordings must be transmitted back to the data center in real-time and processed in conjunction with the permanent seismic network.

The objective of the temporary monitoring network is to create a specific spatial acquisition density and an improved azimuth enclosure for the aftershock zone. It is suggested that there be at least 1 seismic monitoring station within 10 km of the epicenter and at least 4 seismic monitoring stations within 30 km. For earthquakes of magnitude 7 or above, additional seismic stations are necessary on both sides of the ruptured fault. The augmented network should provide rather equal azimuth coverage within 100 km of the epicenter.

The site selection and station building should be in accordance with the *Technical regulations of China digital seismic network* (JSGC-01), although in an emergency case, this is optional. The observation should be situated in a location with stable geological conditions, good communication conditions and low levels of background noise and interference to assure the instrument's safety and functionality.

Observation equipment must utilize broadband or short-period seismographs with low self-noise, high sensitivity and good consistency in accordance with the *Technical requirements of instruments in network for earthquake monitoring seismograph* (DB/T 22-2020). The instrument's time will be synchronized with GNSS, and the timing precision should be less than 1 ms. Dedicated or public wireless or wired links, such as satellite,4G/5G wireless or FTTH network, can be used for the transfer of real-time data, depending on the site conditions and transmission bandwidth and stability requirements. The power is provided by solar energy or AC electricity, with a backup battery. The power capacity should be sufficient for both continuous data gathering and transmission in real-time. The observation system must have adequate lightning and water protection.

During the operation of the temporary monitoring network, the health status of the instruments and communication links, as well as the data quality, should be monitored on a frequent basis so that instrument faults or link failures can be detected and repaired in a timely manner.

In order to record the seismic sequence more completely, it is advised to finalize the field installation within one to two days after the major quake. Additionally, it is recommended to set the sampling frequency at or above 100 Hz and to maintain observations for a minimum duration of three months.

2. Dense seismic array

The primary goal of the dense seismic array is to image the fault or source region's fine structure. Consequently, the seismogenic fault will serve as the observational focal point, and sensors will be installed densely on both sides of the fault to roughly cover the source region and the perimeter.

Observation instruments are advised to utilize readily deployable integrated seismographs or node seismographs. All instruments are time-synchronized using GNSS to an accuracy of 1 ms or better. The array stations are deployed unequally (meters to several kilometers) along multiple cross-fault profiles or scattering in the source zone according to certain rules. To record the wave fields associated with a fault zone, the observations close to the fault zone should always be intensified. Due to the need to build a large number of stations in a short amount of time, the standards for site selection and instrument installation can be lowered or streamlined. As the stations will mostly rely on internal or external batteries for power supply, the battery capacity must be selected with care to provide the expected duration of observation. As the majority of stations are unattended, it is necessary to include safety measures such as warning signs.

If black (unused) fibers are available in the focal area or nearby, fiber-optic-based distributed acoustic sensing (DAS) is an alternative method for dense observation. DAS has developed as a low-cost, simple-to-deploy and low- maintenance technology for high-resolution seismic sensing. Several tens of kilometers can be monitored by a single DAS interrogator unit in meter intervals. Consequently, the fiber sensing network comprised of one or more DAS units can be a valuable addition to the conventional dense seismic array.

The recommended dense observation period is at least one month, and the sampling frequency should not be less than 100 Hz. The field work should be completed within 5–10 days after the major earthquake.

3. Earthquake sequence study

Once the temporary seismic network is operational, more little earthquakes will be detectable. It is necessary to systematically re-scan the waveforms combining permanent and temporary seismic stations in and around the focal zone with advanced automatic detection methods, such as the waveform matched filter technique or deep learning detection, to rebuild a more complete earthquake catalog independent of newly detected events or by merging it with the network center's catalog. The locations and magnitudes of the events can be estimated based on the re-picked arrival times and amplitudes of P- and S-waves. Additionally, the new catalog's quality, magnitude completeness, detection performance and phase-picking performance should be evaluated.

The re-picked phase data could be classified based on their waveform qualities (e.g., SNR etc.) and will be utilized to invert 1D or 3D velocity structures in the focal region. Both will be used to improve the absolute or relative locations of seismic events with additional information regarding travel-time differences from the phase time or waveform correlation of station and/or earthquake pairs.

With a more dense seismic network, the source parameters of more earthquakes can be retrieved, providing more opportunities for sequence research. These parameters, such as the focal mechanism solution and rupture direction, can be inverted from the waveforms with the green function library constructed from updated 1D or 3D velocity structures, and/or with the polarities of the first arrival phase and P/S amplitude ratios extracted from the high SNR seismography, using either conventional methods or a deep learning approach. In addition, the uncertainty of these parameters would be evaluated based on their fitting errors.

The characteristics of the earthquake sequence will be systematically analyzed based on their foreshock migration, the spatial–temporal seismicity pattern of the entire sequence, and the stress evolution derived from b values and source parameters, which will provide insight into the mechanism of earthquake initiation.

On the basis of the Coulomb stress changes of a strong earthquake and the migration and spatial distribution of aftershocks, the triggering and expanding mechanisms of aftershocks will also be discussed.

4. Deep seismogenic environment research

The 3D P- and S-wave 3D velocity structures and event locations surrounding the focal zone will be inverted simultaneously with phase and waveform data from all available permanent and temporary stations. The inversion should adapt the model grids to the distributions of earthquakes and stations, and the updated 1D velocity structure can be used as the initial model. The resolution and uncertainty of the inversion should be examined.

By integrating the inversion of the travel times of body waves with dispersion curves of Rayleigh or Love waves, the resolution and reliability of shallow 3D velocity structures can be enhanced further. In general, dispersion curves can be derived from ambient noise cross-correlation for station pairs consisting of both permanent and temporary stations.

On the basis of the 3D P- and S-wave velocity structures, 3D attenuation structures for P- and S-waves (Q_P and Q_S) may also be inverted using attenuation parameters for events on their seismic rays, which are retrieved from the amplitudes of body waves or coda waves. The resolution and uncertainty of the inversion should be examined. It is possible to jointly invert the 3D attenuation and velocity structures, as well as the event locations.

The 2D or 3D spatial distributions of the three principal stress directions and the stress ratio (R), which are indications of the background seismogenic stress field, will be inverted using the focal mechanism solutions from this earthquake sequence and past large earthquakes.

If there is sufficient data support and a need for scientific investigation, other seismological methods can be used to refine the 3D structures of the media and stress in the focal area, e.g., constraining the diversities of stress or structure under the station with the seismic anisotropy; or to explore the dynamic changes of the underground medium or stress, e.g., revealing the co-seismic medium velocity changes or fault healing based on the repeated earthquakes and ambient noise cross-correlation.

On the basis of the aforementioned results in seismic locations, 3D medium, and stress structures, a fine 3D seismogenic model of the focus zone will be developed. Based on the shape and structural properties of seismogenic faults, as well as the deep seismogenic medium and stress environments, the dynamic process underlying the occurrence of strong earthquakes will be examined.

13.4 Procedures

The working procedure is illustrated in Fig. 13.1.

1. Indoor preparation

The installation sites of temporary seismic stations for both the aftershock monitoring network and the dense array should be preselected and their routes should be plotted in advance using a geographic information system or map. The field work preparations include charging the internal or external batteries of instruments, testing seismographs, solar power suits, and wireless data transmission devices, and gathering tools and materials for field station construction, among other tasks.

2. Field installation

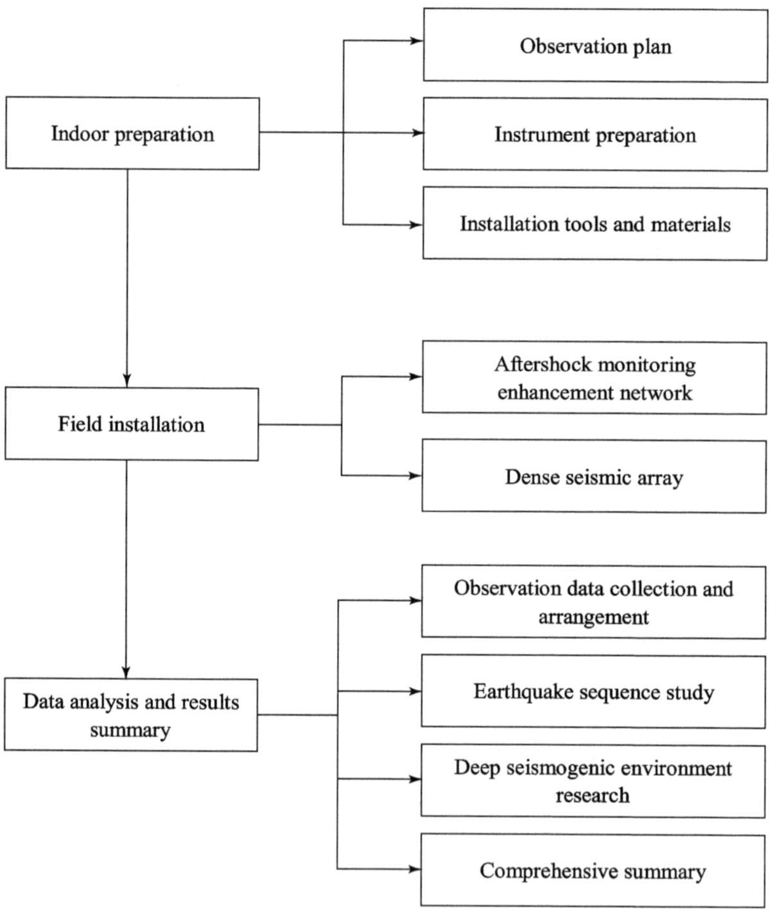

Fig. 13.1 The workflow of the temporary seismic array

Priority should be given to the establishment of a temporary seismic network to improve the monitoring of aftershocks. In general, broadband or short-period seismometers with high sensitivity will be utilized in this configuration. A suitable location for a seismic station will be chosen in the field around the selected point on the map. The seismograph, solar power supply system, and wireless data transmission system will be installed in accordance with the applicable technical regulations. After adjusting parameters and repositioning the seismometer mass, the seismometer, digitizer, and real-time transmission will be tested and safeguarded with water- and heat-resistant materials. The installation will be photographed, and metadata regarding the station's coordinates and instruments will be recorded. The stations are regularly tended by adjacent residents.

The dense seismic array will be implemented concurrently or shortly after the aftershock monitoring network. If the same traditional equipment is used in a dense seismic array, the configuration will be identical to the previous network. However, the installation will be greatly simplified through the use of all-in-one short-period seismography or node seismography, which will dig a hole in a relatively quiet and secure location near the map-selected place and bury the seismography and external batteries inside. Prior to burial, the condition of GNSS time locking, three-component waveform, data writing, etc., should be evaluated separately. Additionally, safety measures such as water resistance and a warning sign should be implemented. After collecting metadata and taking photographs of the surrounding area, the installation of the unmanned station will be complete. If distributed acoustic sensing is to be applied to the dense array, the available fiber resources must be arranged into one or more long-distance pass-through fiber routines. The DAS interrogator unit (IU) will be installed and connected at one end of a fiber route. The maximum distance that a DAS can detect is contingent on the DAS IU and the available fiber supply. Once the parameters are set, the seismic array with a high density along the wire is ready to function. The sensing locations must be positioned throughout the fiber using active methods, such as hammering.

To intensively record aftershock activity for a complete investigation, all stations must be deployed as quickly as feasible following the primary earthquake. To expedite the field work, it is recommended to set up the instruments with multiple working groups.

3. Data analysis and results summary

(1) Collection and analysis of field observation data. Included in this collection are station metadata, waveform data, and log files. The meta primarily provides the station coordinates (latitude, longitude, and elevation), instrument type, and transfer function of the seismography, among other information. Conforming to the *Seismic waveform data exchange format* (DB/T 2-2003) specification, the waveform data are continuous and in MINISEED or SEED format.

New observation data is analyzed in order to advance seismic sequence and seismogenic environment studies. These include, but are not limited to, earthquake detection and precise localization, calculation parameters for seismic source and seismicity, inversion of three-dimensional velocity and attenuation structures, etc. Methods for processing data are numerous, and both conventional and modern seismology techniques can be utilized. For routine processing, mature and widely used software programs are typically employed. It is proposed that different research organizations can utilize different mature approaches to process the same data and then compare their results to determine the differences and possible ways to improve the reliability of the results.

(2) Application of modern seismological methods. Many seismology applications have gained new vitality, such as the use of deep learning for the automatic identification of seismic phases and events, and in many cases, its recognition accuracy and Recognition consistency have surpassed human manual recognition, significantly reducing the intensity of human labor and enhancing or expanding related applications. On the basis of traditional seismological analysis results, it is proposed to introduce new seismological methodologies, improve conventional applications, expand new applications, and fully mine huge seismic waveform data.
(3) Comprehensive analysis. After summarizing the conventional or new method processing results of various research groups, a multi-level and multi-angle comparison is conducted, and different results are selected or integrated as the final result for comprehensive interpretation, as well as a deep structure model and seismic activity model are developed.

13.5 Results and Data Archiving

1. Contents and formats

The results of the temporary seismic array observations will include: high-quality seismic waveform data (The waveform continuity rate is not less than 95% and station non-failure rate is not less than 90%); a more comprehensive and precisely located earthquake catalog; focal mechanism solutions for earthquakes of magnitude $M3.5/M3.0$ and above; and refined 3D velocity structures (V_P, V_S, V_P/V_S) and attenuation structures(Q_P, Q_S) in the focal area; stress field characteristics (distributions of stress primary direction and stress ratio R-value); other seismological results. In the summary report, all results will be presented as text, binary data, and individual or comprehensive graphics.

2. The main contents in the summary report

The report should contain, but not be limited to, the following sections:

1) Preface.

This section will comprise the objectives, scope, workload, investigative techniques, field deployment, list of members or working group members, and job assignments.

2) Historical seismological studies.

(1) Distributions of earthquakes in the past.
(2) Characteristics of earthquake activity in the past.
(3) Previous research on regional subsurface structure.
(4) Prior research on the regional stress field.

3) Construction of a temporary seismic array.

(1) Design of an enhanced network for aftershock monitoring.
(2) Design of dense seismic array.
(3) Site selection, installation of stations, and acquisition of data.
(4) Evaluation of the quality of seismic observation data.

4) Research on earthquake sequence.

(1) A more comprehensive earthquake catalog.
(2) Absolute and relative earthquake location refinement.
(3) Solutions for seismic focal mechanisms.
(4) Comprehensive earthquake sequence analysis.

5) Deep seismogenic settings.

(1) 3D velocity structures.
(2) 3D attenuation structures.
(3) Characteristics of the 2D or 3D stress field.
(4) Additional seismic results.
(5) Comprehensive analysis of the seismic environment at depth.

6) Overview and issues.

Make a complete overview of the temporary seismic array, compare the results to the anticipated objectives, and identify the existing problems and weaknesses as a reference for future scientific inquiry.

3. Collection and submission of raw data

Raw or sorted continuous waveform data, station metadata, logs of instrument running and status of health, and logs of the station running from site selection, construction, inspection, and end of running must be collected and submitted.

Chapter 14
Mobile Geomagnetic Observation

14.1 Objectives

Complete one or several phases of repeated geomagnetic field surveys in the earthquake epicenter and the adjacent areas. The observation elements may be geomagnetic vectors (total intensity, magnetic declination, magnetic inclination) or only total geomagnetic intensity according to the specific scheme. According to the obtained geomagnetic field measured results, the spatial structure of the lithospheric magnetic field in the earthquake epicenter and the adjacent areas can be obtained. The temporal variation of the lithospheric magnetic field can be obtained with multi-phase measurements.

14.2 Contents

According to the nature of the earthquake and its damage, the post-earthquake scientific research plan is formulated, which will determine the objectives, scale and elements. According to the confirmed scientific research plan, combined with the existing geomagnetic measured data in the earthquake area, complete one or more of the following work contents.

1. Intensive monitoring of abnormal variation of the lithospheric magnetic field vector

The temporal variation characteristics of the lithospheric magnetic field before and after earthquakes can be studied by means of the geomagnetic field vector survey conducted within a radius of no less than 300 km around the epicenter and the geological structure of interest, combined with the previous geomagnetic field vector survey data in this area. The elements being studied include the temporal variation of the

Institute of Earthquake Forecasting, China Earthquake Administration,
A Guidebook to Earthquake Scientific Investigation,
https://doi.org/10.1007/978-981-96-0418-0_14

vertical component, the horizontal vector, the east component, the north component and the total intensity.

2. Lithospheric magnetic fine structure detection

The spatial structure of the geomagnetic field in the area of concern was explored through high-resolution geomagnetic total intensity detection at the epicenter and seismogenic fault. And the spatial distribution of the three-dimensional lithospheric magnetic field can be obtained by combining the aerial magnetic data. The lithospheric magnetic field is further separated layer by layer to obtain the depth distribution inside the lithosphere and the spatial distribution characteristics of the lithospheric magnetic field at different depths. Finally, the spatial three-dimensional magnetic structure of the lithosphere is obtained.

3. Magnetic susceptibility structure detection across fault profiles

By laying the total geomagnetic intensity detection profile across the focus fault, the underground magnetic susceptibility structure of the profile is analyzed, and the medium structure of the focus fault is detected. By stripping the lithospheric magnetic field layer by layer, we can invert the interior depth of the curie surface and magnetic susceptibility distribution of the profile, which can provide magnetic proof for the fault structure.

14.3 Methods and Basic Requirements

1. Intensive monitoring of abnormal variation of the lithospheric magnetic field vector

1) Field work.

The geomagnetic elements of the abnormal variation of the lithosphere magnetic field includes the total geomagnetic intensity, magnetic declination and magnetic inclination. Each measure team will be equipped with three proton precession magnetometers, two fluxgate theodolites and three GNSS instruments. According to the number of measurement stations, it is necessary to deploy multiple teams to carry out field work simultaneously, so as to complete the intensive monitoring as soon as possible. The average spacing of the measuring stations is about 70 km, and the spacing can be increased appropriately in the western region. The spatial position of the original measuring stations shall be used as much as possible. The measurement process is in conformity with *basic technical requirements for roving geomagnetic measurement (trial)*.

2) Data processing.

(1) Diurnal variation reducing.

The main purpose of diurnal variation reducing is to eliminate the regular diurnal variation field and other external field components in the measured geomagnetic data and to obtain the internal field in the measured geomagnetic data. The following two points should be paid attention to in the process of daily variation reducing:

① When the observatory is far away from the measuring station, it should be considered to set up some continuous magnetic field measurement instruments near the monitoring station to reduce the errors caused by the spatial unevenness of the diurnal variable of the geomagnetic field.
② The choice of a target date in diurnal variation reducing should be very careful, usually a magnetic quiet day near the measure date should be selected as the target day. According to experience, the distance between the measure date and the target date should not exceed 30 days.

(2) Secular variation reducing.

The monitoring target of intensive monitoring of abnormal variation of lithospheric magnetic field vector is to describe the spatial distribution and temporal variation of the lithosphere magnetic field, that is, the difference between the lithosphere magnetic field of the current period and the previous period. It is necessary to calculate and eliminate the secular variation of the main field in the diurnal variation reducing the results of the two periods according to a secular variation model of the geomagnetic field. The available geomagnetic field model can be the International Geomagnetic Reference Field (IGRF) or a secular variation model of the geomagnetic fields in China by the Natural Orthogonal Component (NOC) method.

3) Analysis of anomalies.

The results of the intensive monitoring of abnormal variation of lithospheric magnetic field vector are the temporal-spatial variation and its abnormal characteristics in the epicenter area. The geomagnetic elements analyzed include total geomagnetic intensity, magnetic declination, magnetic inclination, north component, east component, vertical component, horizontal component and horizontal vector (H_y, H_x).

2. Lithospheric magnetic fine structure detection

1) Field work.

The geomagnetic element of lithospheric magnetic fine structure detection is the total geomagnetic intensity. Each measure team will be equipped with three proton precession magnetometers. According to the number of measurement stations, it is necessary to deploy multiple teams to carry out field work simultaneously, so as to complete the intensive monitoring as soon as possible. The average spacing of the measuring stations is about 5 km, and the best reduction is 1–2 km. And keep the lattice layout rule, each side of the measurement station is not less than 10 points is appropriate. The measurement process is in conformity with basic technical requirements for roving geomagnetic measurement (trial).

2) Data processing.

(1) Diurnal variation reducing.

The main purpose of diurnal variation reduction is to eliminate the regular diurnal variation field and other external field components in the measured geomagnetic data and to obtain the internal field in the measured geomagnetic data. The following two points should be paid attention to in the process of daily variation reduction:

① When the observatory is far away from the measuring station, it should be considered to set up some continuous magnetic field measurement instruments near the monitoring station to reduce the errors caused by the spatial unevenness of the diurnal variable of the geomagnetic field.
② The choice of a target date in diurnal variation reduction should be very careful, usually a magnetic quiet day near the measure date should be selected as the target day. According to experience, the distance between the measurement date and the target date should not exceed 30 days.

(2) Main magnetic field removing.

The main purpose of the main magnetic field removal is to eliminate the main magnetic field in the diurnal variation reducing results and obtaining the lithospheric magnetic field in the measured data. International Geomagnetic Reference Field (IGRF) is a usually used reference geomagnetic model of the earth's main magnetic field in the observation area. After subtracting the values from the main magnetic field model with the same spatial positions from the results of diurnal variation, the main magnetic field parts in the geomagnetic internal source field were eliminated, and the spatial distribution data model of the lithosphere magnetic field was obtained.

3) Data parsing.

The stratified matching filter method is used to analyze the depth and shallow partial separation of the lithospheric magnetic field. It is generally believed that the magnetic field anomalies generated by shallow geological bodies are sharper than those generated by deep geological bodies. Therefore, the magnetic field anomalies caused by shallow geological bodies are usually narrow and have more high-frequency components. However, the geomagnetic anomalies caused by deep geological bodies are usually wider and with a lower frequency end spectrum of slow attenuation. Therefore, the stratified characteristics of the energy spectrum can be used to separate the lithospheric magnetic field spectrum to obtain the shallow field and deep field. The calculation steps are as follows:

① Transform the spatial distribution of lithosphere magnetic field that is into the frequency domain by two-dimensional Fourier transform;
② Calculate the radial logarithmic energy spectrum from the real and imaginary parts of the frequency domain;
③ Construct the matching filter factor according to the logarithmic energy spectrum curve;

④ Multiply the frequency spectrum of the lithosphere magnetic field by the corresponding filtering factor to obtain the spectrum of the shallow source field and the deep source field;
⑤ Apply the inverse Fourier transform from the two parts of the spectrum to obtain the separated shallow source field and deep source field.

4) Curie isothermal surface inversion.

The depth and undulation pattern of Curie isothermal surface can be obtained from deep lithosphere compensated inversion. The principle is that the Curie isothermal surface can be regarded as a transverse continuous model, so Curie isothermal surface can be converted into an average grid center point with a certain distance, and each grid center point represents the depth of the grid of the Curie isothermal surface. Curie surface inversion is under the following two assumptions:

① The intensity of magnetization of the magnetic shell is uniform and the thickness of the magnetic can be changed;
② Since the fluctuation of the earth's surface is generally much smaller than that of the Curie surface, the fluctuation of the magnetic upper interface (the earth's surface) is ignored.

Using the approximate linear inversion formula, the fluctuations of the Curie surface with respect to the average depth are expressed as finite Fourier series, and the Curie depth is obtained by the least square method.

3. Magnetic susceptibility structure detection across fault profiles

1) Field work.

The geomagnetic element of magnetic susceptibility structure detection across fault profiles is the total geomagnetic intensity. Each measure team will be equipped with three proton precession magnetometers. According to the number of measurement stations, it is necessary to deploy multiple teams to carry out field work simultaneously, so as to complete the intensive monitoring as soon as possible. The profile should be as vertical as possible and focus on the strike of the fault and extend at least 50 km on both sides with the fault as the center. The average spacing of the measuring stations is about 5 km, and the best reduction is 1–2 km. It is suggested that at least two geomagnetic profiles should be laid at different positions of faults at the same time. The measurement process is in conformity with *basic technical requirements for roving geomagnetic measurement (trial)*.

2) Data processing.

Please refer to the previous relevant content.

3) Magnetic susceptibility inversion.

Lithospheric magnetic anomalies are caused by underground magnetic bodies under the action of the geomagnetic field, and the underground medium is continuously distributed, which can be seen as a series of magnetic anomalies composed of high or low magnetic susceptibility blocks. When the distribution of underground magnetic susceptibility is retrieved from the lithospheric magnetic anomalies observed on the ground, the underground space medium in the observed profile can be divided into a series of block units. When the block elements are small enough, it is assumed that the magnetic susceptibility is evenly distributed within each block element, and the differences in magnetic susceptibility are manifested as differences between different block elements. According to the potential field theory, when the influence of remanence and demagnetization is not taken into account, the magnetic field generated by a block with uniform magnetic susceptibility is proportional to the magnetic susceptibility, and the proportionality coefficient is only related to the shape and spatial coordinates of the block element. When the block element size and position are fixed, the proportional coefficient can be determined in advance. In the inversion process, the magnetic susceptibility value is the only parameter to describe the geological block, and also the only parameter to be inverted. Thus, the problem of retrieving the source body of the field from the lithospheric magnetic anomalies becomes the subsurface distribution of the magnetic susceptibility of the block units.

14.4 Procedures

The operation procedure of the roving geomagnetic survey is shown in Fig. 14.1.

1. Indoor preparation

1) Document preparation.

(1) Collect the parameters of the earthquake.
(2) Collect geomagnetic data related to the earthquake area, including regional aeromagnetic data, EMAG2 data, historical geomagnetic data, etc.
(3) Collect different scale topographic maps, geological maps, traffic maps, etc., related to the earthquake area.
(4) Collect high-resolution satellite and aerial images related to the earthquake area and preliminarily interpreted them.
(5) Formulate team organization, content, focus and routing for field work.
(6) It is suggested that collect and sort out data and materials in dangerous areas and even in Chinese nationwide in daily work if condition permits. In order to prepare for the post-earthquake scientific work can quickly obtain the relevant data and materials.

2) Instruments and equipment.

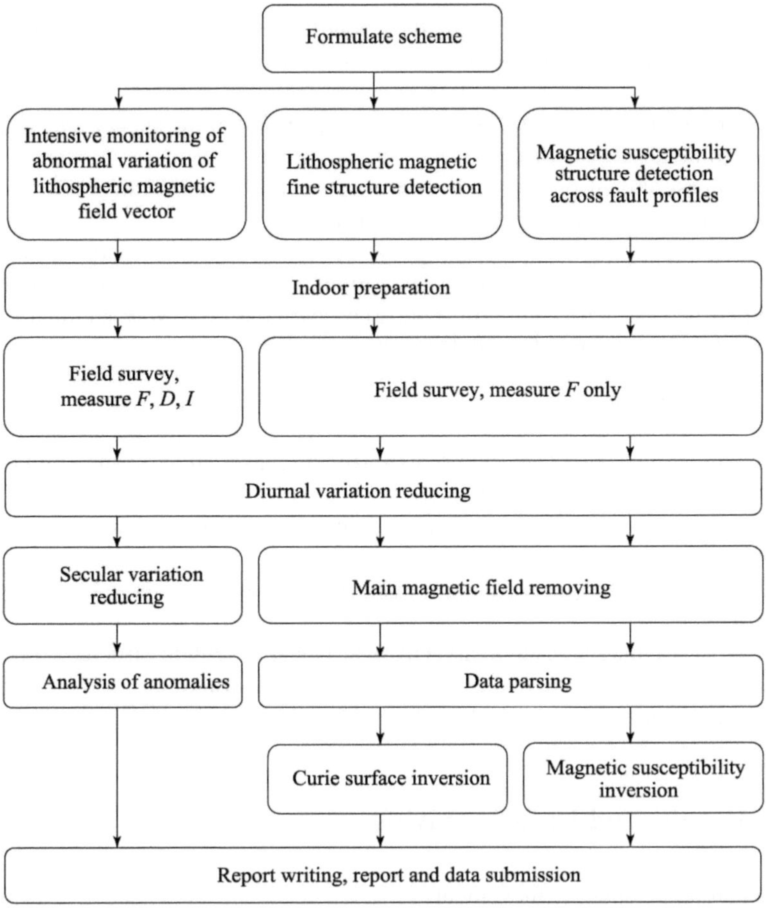

Fig. 14.1 Operation procedure of roving geomagnetic survey

(1) All kinds of field measurement instruments and tools. It mainly includes fluxgate theodolite, proton nuclear meters, GNSS instruments, handheld GNSS, geological compass, laser rangefinder, tape measure, ice axe, shovel, drone, satellite phone, spray paint and so on.
(2) Prepare all kinds of non-magnetic field work equipment and labor protection articles.

2. Field survey

According to the general requirements of the scientific research scheme, this guide should be applied to carry out the corresponding geomagnetic field survey.

(1) Due to the difficult working conditions, urgent time request and heavy tasks, the field work should be divided into several teams. The teams should strengthen

communication and comply with the scientific research plan in accordance with the unified working methods and technical ideas on the premise of ensuring safety.

(2) In field work, each team should start from one side of the survey area together, and gradually a systematic advance in one direction.
(3) The measurement process is in conformity with basic technical requirements for roving geomagnetic measurement (trial).
(4) The field data should be carefully checked on the spot, the data measured on that day must be collated on the same night, and the progress of the field work should be timely uploaded. In order to prepare the field work summary report quickly after the end of the field work.

3. Analysis and summary

(1) Download the time series data of geomagnetic observatories, and conduct data verification and preprocessing.
(2) Diurnal variation reducing, secular variation reducing, main magnetic field removal and other data processing processes.
(3) Time variation, frequency domain analysis, inversion analysis and other analysis of the data.
(4) Analysis of magnetic anomaly characteristics, magnetic structure, magnetic susceptibility structure and underground medium structure.
(5) Roving geomagnetic research summary report writing.

14.5 Results and Data Archiving

1. Content and form of achievements

The achievements of the roving geomagnetic survey are mainly reflected in the location of the earthquake epicenter, the seismogenic fault, the fault of concern, the spatial distribution characteristics of the lithospheric magnetic field and the magnetic susceptibility structure around the geological body. After the field work, the final summary report with pictures and related attachments including related achievements shall be submitted.

2. Main contents of the summary report

The report shall include but not be limited to the following main contents:

1) Introduction.

This part includes the basic parameters of the earthquake, the objective of the geomagnetic survey, a brief description of the content and workload of the survey, the participating units, etc.

2) Field work scheme.

This part includes the distribution of measuring stations, the missions of each participating unit, the field survey date, the number of measurement staff, etc.

3) Data processing.

This part includes data preprocessing, diurnal variation reducing, and secular variation reducing shall be included if the scientific research content is "intensive monitoring of abnormal variation of lithospheric magnetic field vector"; and main magnetic field removing, data parsing and curie isothermal surface inversion shall be included if the scientific research content is "lithospheric magnetic fine structure detection"; and magnetic susceptibility inversion shall be included if the scientific research content is "magnetic susceptibility structure detection across fault profiles".

4) Analysis of the characteristics of lithospheric magnetic field characteristics.

The spatial distribution characteristics of each element of the lithosphere magnetic field and magnetic susceptibility, including the total intensity anomalies of the lithosphere magnetic field. If the research content is "intensive monitoring of abnormal variation of lithospheric magnetic field vector", it also includes magnetic declination, magnetic inclination, north component, east component, vertical component, horizontal component and horizontal vector (H_y, H_x), etc.; if the research content is "lithospheric magnetic fine structure detection", it also includes the deep and the shallow structures of the total intensity, Curie surface depth, etc.; if the research content is "magnetic susceptibility structure detection across fault profiles", the distribution of magnetic susceptibility is also included.

5) Analysis of the underground structure.

This part includes the temporal and spatial distribution characteristics of earthquakes in the research area and the significant abnormal characteristics of the magnetic field. If the research content is "lithospheric magnetic fine structure detection", it also includes the distribution of Curie surface, if the research content is "magnetic susceptibility structure detection across fault profiles", the structural characteristics of magnetic susceptibility are also included.

6) Problems and deficiencies.

This part makes a comprehensive summary of the work of the roving geomagnetic research, compares the results of the work with the expected objectives, puts forward the existing problems and shortcomings and provides a reference for future research work.

3. Data submission

The original data submitted includes but is not limited to the following:

(1) Involved geological, mapping and geographic information in the roving geomagnetic survey.

(2) All kinds of original measured data obtained in roving geomagnetic field work, including the total intensity of the geomagnetic field, longitude and latitude data of measuring stations, etc. If the scientific research content is "intensive monitoring of abnormal variation of lithospheric magnetic field vector", magnetic declination and magnetic inclination shall be included.
(3) All kinds of photos, videos, and text descriptions in the roving geomagnetic research.
(4) Confidential information shall be filed and managed in accordance with relevant security management measures.

Chapter 15
Strong Ground Motion and Engineering Earthquake Damage Investigation

15.1 Objectives

Collecting and processing the observed strong ground motion records of the earthquake sequence, using the observed strong ground motion records and the rapid intensity reporting station records, analyzing the time–frequency characteristics of strong ground motion, and combining with observed records thus to provide the ground motion influence field; developing earthquake damage survey of engineering structures, estimating the damage grade and characteristics of engineering structures, conducting aftershock monitoring for typical engineering structures and investigating casualty factors through site survey to provide more data for the improvement research of the seismic performance of buildings and the rapid estimation of earthquake damage.

15.2 Contents

1. Strong ground motion

Collecting and processing the observed strong ground motion records, analyzing time–frequency characteristics of ground motion and its correlation with earthquake damage, combining with ground motion simulation methods to provide the spatial distribution of ground motion intensity measures and conducting aftershock monitoring for engineering structures.

1) Strong ground motion data.

Collecting the observed strong ground motion records and the earthquake intensity rapid reporting network records, obtaining information about earthquake hypocenter, path and site condition of records. Processing the observed records with regular

Institute of Earthquake Forecasting, China Earthquake Administration, *A Guidebook to Earthquake Scientific Investigation*, https://doi.org/10.1007/978-981-96-0418-0_15

approaches, acquiring the peak ground acceleration, the peak ground velocity, the Fourier amplitude spectrum and the (pseudo) acceleration response spectrum of the records. Comparing the acceleration response spectrum of observed records from near-field stations and important areas with the design response spectrum and analyzing time–frequency characteristics of ground motion and its correlation with engineering damage.

2) Simulation of the ground motion field.

Utilizing the earthquake hypocenter rupture model, based on the stochastic finite fault ground motion simulation approach, considering the attenuation on the propagation path in the earthquake-prone areas as well as the regional characteristics of local site effects. With the calibration of observed strong ground motion records, simulating the acceleration time history of ground motion in the hypothetical intensive observed sites within a certain region of the epicenter, providing the peak ground acceleration, peak ground velocity, acceleration response spectrum and other ground motion intensity measures.

3) Determination of the ground motion influences the field.

Providing the amplitude of ground motion and the spatial distribution of spectral acceleration response at certain periods, confirming the influence region and the key influence areas of earthquakes, based on the observed and simulated ground motion records.

4) Geological information collection of the strong ground motion stations.

Collecting the site condition data of the triggered strong ground motion stations, especially the borehole data of the sites, and considering the damage grade of the surrounding buildings to analyze the characteristics of the strong ground motion records. For those stations without borehole data, site exploration could be done according to practical requirements, getting the site condition data through shallow soil drilling work.

5) Characteristic testing of engineering structures and aftershock observation.

Setting up strong motion stations near the epicenter or within the typical engineering structures for microtremor and aftershock observation, and determining the dynamic response performance of engineering structures, to apply for the earthquake resilience estimation of engineering structures.

2. Engineering damage reconnaissance

Conducting earthquake damage reconnaissance of engineering sites, buildings, structures, a major engineering, geotechnical structures, hydraulic structures, underground structures and lifelines. Determining the earthquake damage grade, examining the earthquake intensity results according to the intensity code, investigating the casualty factors and developing the assessment of earthquake resilience comprehensively.

1) Sites and their damage reconnaissance.

The geomorphic units, the geomorphic genetic types, the slope, the relief and some other earth morphological characteristics of the sites. The types, structures, and thickness of the foundation soil on engineering sites, as well as the genetic types, the development degree of joints and fractures, the thickness of soft interlayer and the layout characteristics, the integrity and other features of the soil body, site classification of engineering sites. Damages to sites including ground fissures, subsidence, sand liquefaction and landslide, etc. The depth of groundwater on liquefied sites, the thickness and types of liquefied soil. The soil distribution, quantity and scope of subsidence and the impact of subsidence on engineering in the subsidence areas.

2) Damage reconnaissance of the building and earthquake intensity.

Classifying the buildings in earthquake areas based on their features, the following types of building structures should be focused, for example, the high-rise frame-tube reinforced concrete structures and tube-in-tube structures, the high-rise reinforced concrete shear-wall structures, the high-rise reinforced concrete frame shear-wall structures, the multi-story and high-rise reinforced concrete frame structures, the multi-story brick-concrete structures, the multi-story frame structures, the multi-story brick-wood structures, the multi-story cavity-wall structures, the single-story reinforced concrete buildings, the single-story brick structure buildings, the single-story steel structure buildings, the spacious buildings, the single-story brick-wood buildings, the single-story brick-concrete buildings, the wood frame houses, the adobe houses, the adobe caves, the loess caves, the galleries or rubbles masonry buildings, etc.

Determining the survey sites, collecting basic information about buildings in survey sites about the amount, types, functions, construction time and detailed drawings of typical buildings. Investigating the damage and estimating the damage grade (namely, basically intact, slightly damaged, moderately damaged, severely damaged and destroyed) of buildings. Determining the proportion of damaged buildings at each survey site, investigating the damage degree of different types and functions of buildings, and reviewing the earthquake intensity following intensity code.

Selecting the representative buildings (which illustrate the damaging mechanism of buildings or representative damage form in this intensity area) in different intensity areas, conducting a detailed survey on building damage, taking photographs, and making descriptions of building damage.

3) Damage reconnaissance of structures and major engineering.

Surveying the damage of structures and major engineering, according to the engineering characteristics in earthquake areas, including blast furnaces and trestles of metallurgical enterprises; tanks, towers, mining derricks, industrial or civil chimneys of chemical enterprises; nuclear power plants, offshore oil production platform, large dams, large wharves, airports, transportation hubs, large bridges, television towers and some other engineering structures and important infrastructure.

4) Damage reconnaissance of geotechnical and hydraulic structures and underground engineering.

Investigating the damage of geotechnical structures (soil dams, embankments, retaining walls, etc.), hydraulic structures (sluices, small and medium-sized flood dams, pumping stations, etc.), recording the direction, width, length and depth of fractures, investigating the earth damage and whether there were foundation slips in the nearby areas; investigating the damage of mines, underground shopping malls and civil air-defense engineering of underground engineering, writing down the fractures and their directions, investigating whether there were faults or passing through the underground engineering.

5) Damage reconnaissance of lifelines.

Investigating the damage of lifelines in different intensity areas, and classifying damage grade, including the water pools, water treatment pools, water pipelines, water towers and others of the water supply system; gas pipelines, gas storage tanks and others of gas supply system; drainage pipelines, water treatment pools and others of drainage system; roads, bridges, tunnels, airports, wharves, culverts and others of the transportation system; power facilities, various of factory buildings, auxiliary engineering facilities, electrical facilities and transmission lines in transformer substation and dispatching communication networks of power supply system; buildings and structures related to telecommunication, communication facilities and transmission lines in a communication system.

6) Earthquake casualty reconnaissance.

Investigating and analyzing casualty locations in this earthquake, casualty features and causes of deaths and injuries; analyzing the correlation between earthquake casualties and the occurrence time of earthquake; investigating the damage of structures and nonstructural components, the damage density, degree and size of geological disasters and the correlation factors with the casualty.

7) Earthquake resilience assessment of engineering system.

Investigating the damage of structures and nonstructural components, and exploring practical resilience technology applied to the recovery and reconstruction of disaster areas; conducting earthquake resilience assessment based on the infrastructure subsystem; investigating the post-earthquake emergency response, transitional settlement, recovery and reconstruction actions of all-level government administration departments and publics, and its effects, combining with the resilience of structures and infrastructure system of lifelines, evaluating the earthquake resilience comprehensively.

15.3 Methods and Basic Requirements

1. Strong ground motion

1) Methods.

Collection of strong ground motion data by telecommunication, data processing and ground motion field simulation, analysis of the time–frequency characteristics of ground motion, mapping of the spatial distribution of strong ground motion intensity measures and the ground motion influence field by office work; survey and collection of the site information of strong ground motion stations; analyzing and selecting typical engineering structures to conduct aftershock observation.

2) Basic requirements.

(1) Examining the data quality of the strong ground motion records, collecting the basic information about hypocenters, paths and sites of strong ground motion records.
(2) Processing the strong ground motion records with regular approaches, the basic procedure should include baseline correction, windowing and zero-adding at the beginning and end of the records, bandpass filtering, velocity and displacement integral and computing various ground motion intensity measures (the peak ground acceleration, the peak ground velocity, the 5% damped pseudo-spectral acceleration response at engineering related periods range and the Fourier amplitude spectrum).
(3) Figuring the acceleration time history, velocity time history and the (pseudo) acceleration response spectrum of observed records from near-field stations and important areas by drawing software.
(4) Comparing the acceleration response spectrum of observed records from near-field stations and important areas with seismic design code response spectrum at different fortification levels.
(5) The simulation results should be compared with the observed records and ensure certain consistency. Vital parameters in simulation input could be adjusted according to the observed records; the accessible time and spatial process of rupture model or the slip distribution inversion model should be firstly selected as the earthquake hypocenter rupture model, then the stochastic rupture model based on the empirical statistical laws. The certain range within the epicenter is determined by magnitude, the range of $M \leq 6.5$ earthquakes is an epicenter distance of 200 km, 300 km for $6.5 < M \leq 7.5$ earthquakes and 500 km for $M > 7.5$ earthquakes; the distance of observed spots in simulation should increase with the distance from observed spots to the rupturing fault, the distance of observed spots in near-fault areas should be less than 5 km; the geomorphic data-based site classification and the estimated value of V_{S30} could be taken as the site conditions of observed sites.
(6) Figuring the spatial distribution of earthquake intensity measures using drawing software (for example, GMT), the distribution scope should be controlled within the largest distance of observed spots, and the ground motion influence field should be provided.
(7) Considering the characteristics of the sites of strong ground motion stations, according to existing relevant regulations, conducting a geological survey of

the engineering sites and drilling boreholes at engineering sites. The depth of boreholes depends on the necessary depth for soil response, meanwhile the in situ shear wave velocity testing, the soil dynamic triaxial testing and the regular soil testing should be conducted.

(8) Aftershock monitoring stations of engineering structures should be set up in the earthquake hazard areas where strong aftershocks are likely to occur, the high-intensity areas, and the epicenter areas near the fault. Giving priority to high population density areas and buildings where earthquakes may cause a large number of casualties; giving priority to the developed areas and major engineering that may suffer from severe economic loss; giving priority to the typical engineering structures so that the obtained strong motion records could be applied extensively; in addition, the station site selection should take into account severe factors including transportation, background noise, communication, power supply and safety, etc.

2. Engineering damage

1) Methods.

Using standards issued by the government and seismic community as the work guide for earthquake damage survey, investigating the earthquake damage for engineering structures and the damage state for different engineering structures at different intensity areas through earthquake field survey and expert evaluation.

Taking the national and seismic standards as the work guide of earthquake damage investigation the earthquake damage of engineering structures is investigated by means of on-site investigation, inspection and expert evaluation, where the damage state of various engineering structures in different intensity areas.

2) Basic requirements.

(1) The sites of the engineering damage survey should be reasonably distributed and representative. For those survey sites in rural areas, taking the natural village as the base unit, and the towns should be divided into several districts for the damage survey, each district should be about 1 km^2. For the epicentral areas and areas with earthquake intensity over VIII, every single natural village should be investigated; a sampling survey could be adopted, however, important residences of townships (towns) shouldn't be ignored; all residences should be investigated for sparsely populated regions.

(2) Buildings within the scope of survey sites should be investigated one after another.

(3) Technical requirements of building damage survey should be met.

① Conducting the survey in the areas with earthquake intensity equal to or higher than VI, and the high-intensity abnormal areas with earthquake intensity lower than VI.

② Investigating the brick-concrete structures following the cracks (vertical, horizontal and oblique) and the developed floor of wall cracks, the effect of the ring beams and construction columns, the difference between prefabricated slabs and cast-in-place slabs in different intensity areas. The correlation between the systematic damage and the roof construction and bearing system for single-story workshops, the damage comparison of large prefabricated slab workshops and the workshops with other types of slabs, the correlation between the damage of enclosure walls and the conjunctions with columns, the damage forms of the small and large columns and the damage of the brick columns and buttress columns; the damage of the walls, the influence of the wood columns or brick columns to the walls and the displacement of the main beams for the adobe houses; the damage of beams, columns, joints, masonry infilled walls, stairs, elevators, slabs, glass curtain walls, adjacent parts of high-rise and low-rise buildings and the auxiliary structures of roofs, the reinforcement of the beams, columns and joints for the reinforced concrete frame structures; the damage of conjunction beams of the frame and the shear walls, the shear walls, beams, columns, joints, masonry infilled walls, stairs, elevators, slabs, glass curtain walls, adjacent parts of high-rise and low-rise buildings and the auxiliary structures of roofs, the reinforcement of the beams, columns and joints for the reinforced concrete frame structures. Comparing the damage characteristics and common features of buildings constructed at different ages, and the damage characteristics of buildings with or without fortification.

The photographs of investigated objects should be taken when conducting field survey. The panoramic and partial photographs showing the typical damage of various types of structures of different damage grade are necessary, the panoramic photographs should show the integral damage state and the locations of the local and integral damage; the photographs of damage phenomenon which are special or which can represent the damage characteristics of this earthquake event should be taken; the photographs of representative damage state for the same buildings at different intensity areas should be taken; the photographs that illustrate the ground fissures, subsidence, liquefaction, landslide and other ground damage should be taken.

Each photograph should include the location where the photograph was taken, the subject name and its affiliation, the date, the brief description, the seismic intensity of place and photographer.

(4) Technical requirements of lifelines damage.

Conducting the damage survey of lifelines in the areas with earthquake intensity equal to or higher than VII, only conducting sampling survey at areas with earthquake intensity lower than VII.

The damage survey of roads should focus on the crack, uplift, fracture and uneven settlement of the pavements, and the fracture and collapse of the shoulders, retaining walls, horsts and slopes.

Beam bridges should be investigated one by one, the focal point is the damage of main bearing members, including the settlement of the embankment, the settlement of piers and abutments, the horizontal displacement of piers, the concrete fractures of the piers, the damage of conjunctions of brackets and beams, the drop of beams, etc.

The fracture and subsidence of the cover slabs of water pools and water treatment pools, the damage of the pool walls, the conjunction parts of pool walls and cover slabs and the conjunction parts of pool walls and base slabs.

For the underground water pipelines, the damage parts can be found through seeking the water leaking points.

For the support-type water tower, the damage of the support and the water tank itself shall be mainly inspected, and for the cylinder-type water tower, the support and the water tank shall be mainly inspected.

The damage parts of gas supply pipelines through gas leaking points can be found.

For the gas storage tanks, the bearing members, rails and connected parts with rails should be mainly inspected.

For the power supply system, the power facilities, workshops, auxiliary engineering building facilities, the broken wires and cables, the brittle fracture of post insulators, the damage of the main transformers and the incline and collapse of transmission towers should be mainly inspected.

(5) Technical requirements of earthquake casualty survey.

Providing the lists that focus on the times, places and the correlation to structure damage of the deaths, which is related to the ground motion maps, injured people survey could be conducted at a certain scope according to specific earthquake influence areas.

The population of disaster areas and its spatial distribution should be calculated, using the population heatmap and other data to analyze the mobility and dynamic distribution change of population, investigate the occurrence time of earthquake, human mobility, the types and seismic performance of buildings and the geological and environmental factors at disaster areas.

(6) Technical requirements of the critical influence factors and the estimation of earthquake resilience.

Investigating the damage of structures and nonstructural components, the post-earthquake emergency response, transitional settlement, recovery and reconstruction actions and its effects, combining with the resilience of structures and infrastructure system of lifelines, estimating the earthquake resilience comprehensively.

15.4 Procedures

Figure 15.1 shows the procedure of strong ground motion and engineering damage survey.

1. Laboratory preparation stage

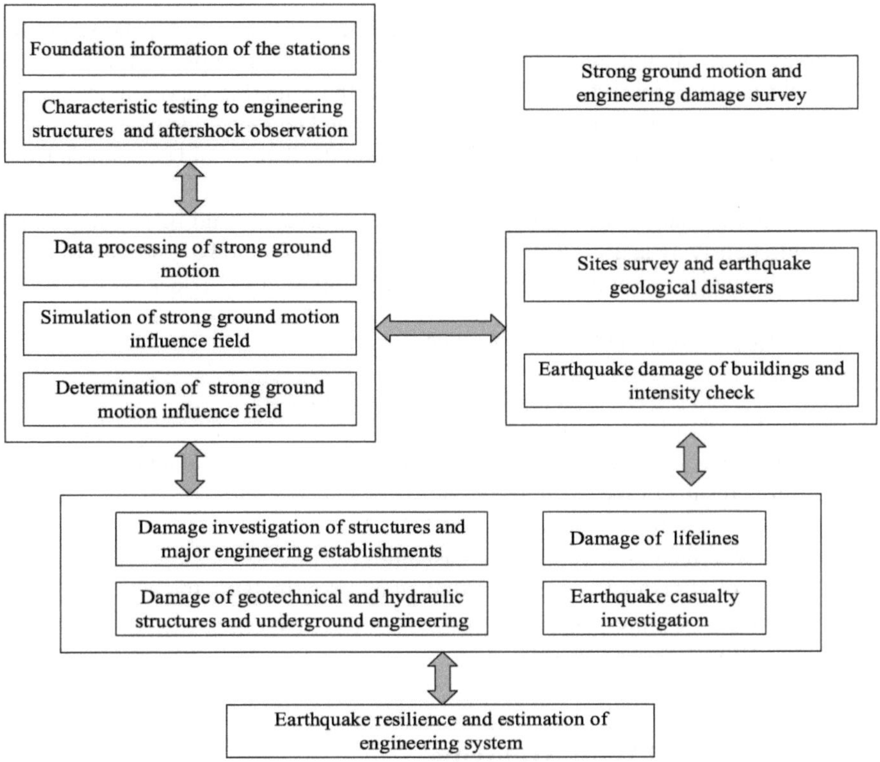

Fig. 15.1 Procedure of strong ground motion and engineering damage survey

1) Data collection.

(1) Collecting basic information of this earthquake event.
(2) Based on the earthquake size and influence, deciding whether to adopt general survey or sample survey.
(3) Selecting the survey spots and determining the types, functions, construction time, detailed drawings and other basic information of engineering structures at the survey spots.
(4) Planning the contents, focal points and routines and other aspects of the earthquake damage survey.

2) Preparation of instruments, facilities and labor insurance supplies.

(1) Various observation instruments and tools of earthquake damage field.
(2) Facilities and labor insurance supplies of earthquake damage field.

2. Field survey stage

According to the general requirements of the scientific survey work, applying this guidance to conduct relevant engineering damage survey.

(1) Owing to the tough conditions, tight deadline and heavy tasks, the scientific survey work should be allocated to severe working groups, communication should be strengthened between groups, and the scientific survey work should be conducted on the basis of unified methods and technical ideas.
(2) Solid work should be done at each survey spot; the data should be abundant and well-recorded.
(3) The working data of earthquake damage field survey should be organized and comprehensively analyzed, and the daily summary should be done carefully, to make it easier to compile the final summary report after the field survey finishes.

3. Laboratory analysis and summary stage

(1) Recovery and processing of strong ground motion data.
(2) Ground motion field simulation.
(3) Characteristics of engineering damage and analysis of typical damage.
(4) Examine results of the earthquake intensity of survey sites.
(5) Characteristics of damage to lifelines.
(6) Summary of factors of earthquake casualties.
(7) Analysis of aftershock monitoring data of engineering structures.
(8) Estimation of engineering resilience.
(9) Compilation of survey reports.

15.5 Results and Data Archiving

1. The content and form of the product

The work product of strong ground motion and engineering damage survey mainly embodies on strong ground motion observed records, the rapid intensity reporting station observed records, the time–frequency characteristics of ground motion, the ground motion influence field, the degree of damage to various buildings in different intensity areas and the characteristics of damage to lifelines. After the survey finishes, the product would be submitted in the forms of survey reports and relevant appendixes.

2. Main content of the summary report

1) Strong ground motion survey.

(1) Acquisition of strong ground motion records.
(2) The observed strong ground motion records time history and comparison of acceleration response spectrum of observed records from near-field stations and important areas with design response spectrum.
(3) Time–frequency characteristics of ground motion and its correlation with earthquake damage.
(4) The scope of ground motion influence field.

(5) Monitoring data from engineering structures and structural earthquake response analysis.

2) Engineering structure damage survey.

(1) Overview of various engineering structures in earthquake areas and characteristics of earthquake damage.
(2) The proportion of different degree of damage to various buildings in different intensity areas.
(3) Typical damage analysis, damage of structures and major engineering establishments earthquake.
(4) Site damage and its influence to engineering structures.
(5) Photographs and their descriptions.

3) Survey of damage to lifelines.

(1) Overview of lifelines in earthquake areas and the impact of earthquake damage on human life and production.
(2) The characteristics of damage to lifelines and the main results of survey.
(3) Comprehensive analysis of the weak link of lifelines and the main factors caused damage.
(4) Necessary drawings, photos and descriptions.

4) Casualty survey.

(1) Casualty lists.
(2) Advices on reducing casualties.
(3) Providing an estimation and modification model for earthquake casualty applied in different areas at different earthquake occurrence time through research.

5) Engineering resilience.

(1) Earthquake resilience analysis of building structures based on the earthquake resilience of the functional engineering system.
(2) Practical advices on resilience technology applied to seismic strengthening and seismic mitigation and isolation engineering.
(3) Earthquake resilience report based on the earthquake resilience of infrastructure subsystem.

3. Raw data submission

The submitted raw data for strong ground motion and engineering damage survey includes but isn't limited to the following contents.

(1) The collected raw data and processed data of observed strong ground motion records.
(2) Ground motion field simulation results.

(3) Aftershock monitoring data from engineering structures.
(4) Typical building damage photographs and descriptions.
(5) Typical lifelines damage photographs and descriptions.
(6) Casualty survey lists.
(7) Estimation of earthquake resilience.

Chapter 16
Seismic Sequence Analysis

16.1 Objectives

To organize a working team quickly after a strong earthquake occurs, who analyze the basic parameters and statistical seismic model of the earthquake sequence, locate the earthquake sequence precisely, inverting the focal mechanisms of the main earthquakes in the sequence and calculate focal parameters such as apparent stress based on the earthquake catalogue and waveform data observed by routine seismic network and mobile observation network. According to the above work result, together with the tectonic background and historical seismic data, the spatial distribution of earthquake sequence and the geometric structure and kinematic properties of the earthquake fault will be provided, then the fracturing fault, earthquake sequence type and earthquake tendency will be determined to provide scientific support for the related work of the earthquake emergency headquarters.

16.2 Contents

1. Analysis of basic parameters from earthquake sequence

Earthquake sequence tracking is a process of time–space scanning and calculating for various parameters by using the monitored earthquake catalogue. For the aim of provide scientific after earthquake predict trend to support the front-line earthquake research personnel determination and rescue. At present, the methods used to judge the sequence type mainly include qualitative analogy analysis and quantitative analysis. Firstly, the regional earthquake type and sequence analogy method are often used to estimate the activity level of the maximum strong aftershock quickly and roughly,

Institute of Earthquake Forecasting, China Earthquake Administration,
A Guidebook to Earthquake Scientific Investigation,
https://doi.org/10.1007/978-981-96-0418-0_16

and then predictive or forecast of the earthquake trend depends on statistical probability. Secondly, continuously analyze and predict the dominant occurrence time of strong aftershocks according to the sequence evolution.

1) Qualitative and analogy analysis.

The type of earthquake sequence and subsequent strong aftershock activity are closely related to geological structure and stress condition. So tracking of earthquake sequence, the geological structure and tectonic environment should be taken as the background, and combined with the distribution of aftershocks, earthquake-generating structure fault feature, fault's location, size, etc., to analyze together (normal fault, reverse fault, strike-slip fault). And then calculate and analyze this earthquake's possible type (main-aftershock, foreshock-main-aftershock, swarm type, isolated type) from the ratio of various types of historical earthquake in the shock region.

Based on the above research, we should analyze the events, which include the epicenter distribution, $M - t$, $N - t$ and cumulative frequency distribution map with magnitude of regional seismic activity, etc., the minimum integrity magnitude level and the relatively stable b-value level, which will represent the seismic monitoring ability of the region, should calculate as the background analysis data for the seismic activity level and spatial distribution characteristics of the study area.

2) Quantitative analysis method.

(1) b-value and b-value intercept analysis of time course.

Based on the minimum complete magnitude and G-R relation, we calculate the b-value. With the change of b-value, the intercept will change. Under a certain number of samples, the magnitude corresponding to the transverse intercept of b-value will be the level of the predicted magnitude in the future. The distribution of maximum aftershock and time analysis.

(2) Based on the statistics of historical earthquake cases in regional, to analyze the characteristics in the focal region by analogy, and obtain the maximum aftershock level in this region, such as 3 days, 7 days, 30 days, 90 days and 1 year.
(3) The experience reference of h value method.

According to the sequence evolution, the h value fluctuates greatly in the first few days and tends to be stable after 4–8 days.

If h value is 0.1–0.3, the most are foreshock type sequence, and h value is above 0.5, the more are main shock-aftershock type, and if h value is 0.3–0.5, the type will be difficult for judgment. If the day by day result of h value changes obviously and is less than 1.0, it perhaps will occur by one or two strong aftershocks, and the dominant corresponding time is usually within 6 days. This criterion has a good effect on predicting the aftershock type sequence of the main shock.

The other methods can be used to analyze type, such as, the waiting time method, the characteristic of significant earthquake cumulative frequency changing with time, the "Segmentation" characteristic of G-R relation, the distribution characteristic of earthquake time interval of earthquake sequence and the comprehensive sequence.

3) Seismic sequence analogy and parameter fitting analysis.

Based on the statistics of historical earthquake cases in the region, the seismic activity, the cases attenuation characteristics, etc., should be analyzed by analogy.

4) The characteristics of seismic activity background and spatial distribution analysis.

According to the Chinese earthquake cases, obtain the seismic activity background and the spatial distribution characteristics, scientifically and accurately to describe the probability occurrence of earthquake. At the same time, the background seismicity intensity can be used to describe the background of seismicity level in the region for a long time.

5) Calculation and analysis of apparent stress.

The apparent stress is calculated by waveform data, and the variation of apparent stress relative to the background value is analyzed. Most of low apparent stress earthquakes belong to main shock-aftershock type. In theory, the higher the apparent stress value indication the higher stress level in the focal area, but the apparent stress is usually related to the magnitude of the earthquake, and the background value related of the magnitude should be deducted.

2. Analysis using statistical seismology models

This study on earthquake sequence using statistical seismology models is mainly separated into two sides for the application of Epidemic Aftershock Type Sequence (ETAS) model. On one hand, it is mainly focused on the obtaining of accurate statistical seismology model parameters using the historical events around the epicenter region. After that, the earthquake sequence can be separated into background sequence and clustering sequence, which represent the long-term background and the ability of triggering their offspring. Then the feature of the spatial and temporal distribution of the sequence can be outlined and we can also determine the probability of each event being a background event. On the other hand, for the aftershocks after a strong event, scientific model parameters could be obtained through the fitting of the earthquake sequence using temporal ETAS model. These accurate model parameters are essential to the significant probability forecasting for the potential of strong aftershocks when the tracking processing will be carried out in the scientific investigation. At the same time, rigorous statistical testing methods should be used to evaluate the performance of forecasting models or algorithms.

1) Basic analysis and model fitting.

(1) Basic analysis for seismicity.

To understand the situation and the spatial–temporal distribution of the earthquakes in study region, the basic analysis containing epicenter distribution plot, magnitude-time plot and cumulative frequency distribution with changes of magnitude should be given. Before the work of using catalogue to statistical seismology models, the completeness magnitude assessing from different views which indicates the ability to monitoring the small events should be calculated.

(2) Fitting of models to the observation.

To significantly simulate the characteristic of the earthquakes in the study region, model fitting based on the assumption from seismology and stochastic study can detect the difference between the models and the observations, especially the model parameters after the process of fitting and its temporal changes from which the rate and the ability to trigger next offspring can be identified.

(3) Analysis for background seismicity and clustering distribution.

The features of total seismicity in a study region can be shown by the background seismicity and clustering events. For example, the statistical declustering algorithm-based spatial–temporal ETAS model can separate the earthquakes into these two sides to analyze the triggering ability and long-term background of the seismicity.

(4) Calculation for probability of triggering next events.

The statistical declustering method based on ETAS model can indicate the triggering contribution from the "parent" events to its next "offspring" events. In general, this algorithm gives a probability value to each event rather than doing this work from a deterministic method.

2) Probability forecasting and performance evaluation.

The probability forecasting is the concept that statistical method should be used to analyze the probability level after the occurrence of strong events. The outputs can consist of the occurrence rate and probability of significant aftershocks, and the performance evaluation using rigorous statistical methods.

(1) Temporal variation of model parameters.

The tracking monitoring on the model parameters which shows the basic features of the earthquake sequence can give the temporal variation of its state. For example, the P value in ETAS model shows the attenuation rate of the aftershock sequence with time after the mainshock.

(2) Forecasting of aftershocks.

After obtaining the accurate and stable model parameters after the fitting of model to the observations, the forecasting of the aftershocks within the windows of certain time, spatial region and magnitude range can be expressed in different time nodes.

(3) Performance evaluation.

To test the performance of the forecasting process of the statistical methods and give the guidance to the scientific investigation, the rigorous test methods should be used to evaluate the consistence between the models assumption and the observations.

3. The fault structure analysis of the seismic sequence

Comprehensive analysis of focal mechanism, precise location results, the time–space evolution characteristics of source-fault structure, fracture model and earthquake sequence are very important. It could provide basic data for strong aftershock prediction of earthquake sequence.

1) The moderate and above moderate earthquake focal mechanism.

The regional seismic waveform inversion method is used to calculate the moderate and above earthquake focal mechanism in the seismic sequence, and the focal mechanism parameters and maps are given for analysis.

2) Precise location of earthquake sequence.

The HYPODD method is regularly used to locate earthquake sequence quickly and the catalogue of earthquake sequence and the maps of epicenter distribution and depth profile are produced. Use the result comprehensive analysis of the focal mechanism, geometric structure, seismic precision positioning and kinematic properties of the faults.

16.3 Methods and Basic Requirements

1. Basic requirements and methods of seismic sequence analysis

1) Determination of the study area.

(1) Methods.

In order to determine the study area, it is necessary to consider the earthquake rupture area and the significant aftershock distribution area comprehensively. If there is obvious cluster activity or significant earthquake event in the marginal zone, it can be analyzed synthetically and independently according to the characteristics of structure and earthquake distribution.

(2) Basic requirements.

Selection of the study area needs to make the number of seismic events involved in the calculation and the background characteristics of seismic activity relatively stable.

2) Parameter selection.

(1) Work method.

The final results of the analysis are greatly influenced by the completeness of seismic catalogue, and the initial parameters are closely related to the characteristics of regional seismicity. Therefore, the selection of the initial parameters should be based on the minimum complete magnitude of the seismic catalogue and the regional seismicity level.

(2) Basic requirements.

The minimum complete magnitude of seismic catalogue should effectively describe the characteristics of regional seismicity, and the parameters can be fitted to a higher degree in the subsequent calculation.

3) The parameter calculation.

(1) Work methods.

The aftershock sequence is calculated and analyzed according to the time node, and the stable parameters can be obtained in time when the sequence approaches the stable state.

(2) Basic requirements.

The parameters representing the sequence characteristics are obtained scientifically and effectively, and the comprehensive reference is provided for the subsequent prediction of the aftershock occurrence trend.

2. Working process and basic requirement of statistical seismology methods

1) Determination of study region.

(1) Working process.

The determination of the study region in statistical seismology methods should consider the rupture area and the covering region of the significant earthquakes. The complement events in a large region can significantly represent the basic activity of the seismicity in the whole study region.

(2) Basic requirement.

The choosing of the study region should ensure the balance between the number of the earthquakes for calculation and the significant events which can indicate the basic feature of the earthquake activity.

2) Choosing of initial model parameters.

(1) Working process.

The initial model parameters which are related to the region feature of the seismicity have a big influence to the fitting result. Thus the choosing of the initial parameters should consider the seismicity level of the study region.

(2) Basic requirement.

The initial model parameters should effectively describe the regional seismic activity characteristics, and the model fitting should be able to achieve effective convergence in the subsequent calculation.

3) Calculation of model parameters.

(1) Working process.

According to the time node of the aftershock sequence tracking calculation and analysis, stable model parameters can be obtained when the earthquake sequence is close to the stable state.

(2) Basic requirement.

The model parameters characterizing the sequence characteristics should be obtained scientifically and effectively, which can be the preparedness for the probability forecasting indicating the occurrence trend of aftershocks.

3. Methods and fundamental requirements of focal mechanism

1) Methods.

According to the seismic clustering characteristics of the earthquake sequence, the research region is selected, and automatic focal mechanism solutions within the research region are collected to give results and check its reliability, which is used as the preliminary focal mechanism solution for earthquake emergency. At the same time, the waveform data of strong earthquakes are collected in the earthquake sequence, and focal mechanism solutions of strong earthquakes in the seismic sequence are manually inverted by the inversion method of regional seismic focal mechanism solution and one-dimensional velocity model.

2) Fundamental requirements.

Focal mechanism solutions of $M \geq 4.0$ aftershocks are inverted for earthquake sequences with mainshocks less than magnitude 7.0. Focal mechanism solutions of $M \geq 4.5$ aftershocks are inverted for earthquake sequences with mainshocks greater than magnitude 7.0. Focal mechanism solutions of $M \geq 5.0$ aftershocks are inverted for earthquake sequences with mainshocks greater than magnitude 8.0. The earthquake is completed on the same day, and the distribution map of focal mechanism solutions for the earthquake sequence is given at the same time.

4. Methods and fundamental requirements of precise earthquake locating

1) Methods.

The seismic phase report of the earthquake sequence is downloaded from the China Earthquake Networks Center, and the form of the seismic phase report is converted to the HypoDD method input format, and the HypoDD method is used for precise locating.

2) Fundamental requirements.

According to the basic terms of scientific research and emergency consultation requirements, the catalogue of precise aftershock locating, the epicenter distribution and depth distribution maps are regularly given.

16.4 Procedures

The working procedure for the study of earthquake sequence is shown in Fig. 16.1.

1. Preparing work indoor

1) Data preparation.

(1) To determine the study region for the basic parameters of statistical parameters of the earthquake sequence according to the cluster characteristics of earthquake sequences before and after the mainshock.

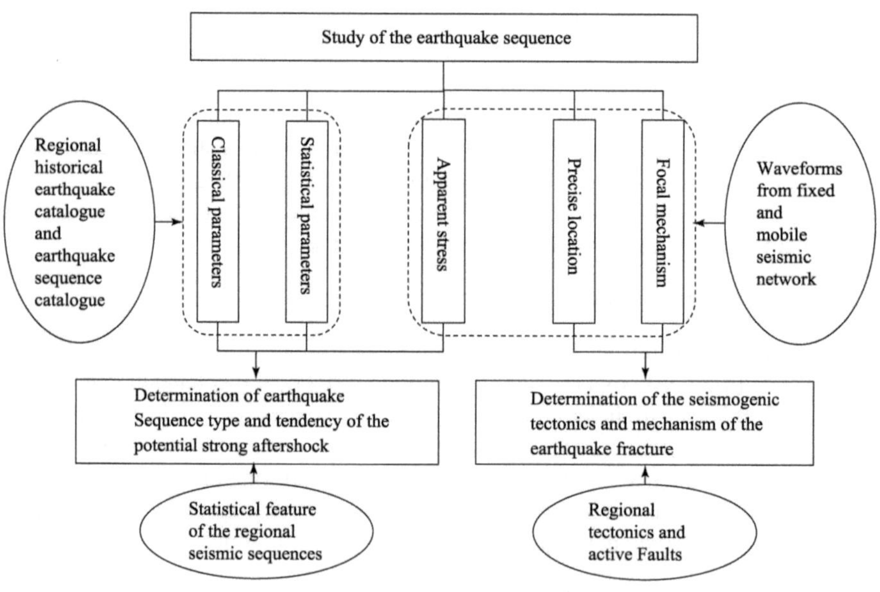

Fig. 16.1 Working procedure for the study of earthquake sequence

(2) To collect various parameters of the earthquake reported by any institutes or research team and determine the study area of the precise seismic location according to the rupture area or region of earthquake sequence activity.
(3) To download the seismic catalogue of the study area from the catalogue data center for the calculation and analysis of the basic parameters and statistical seismic model parameters of the earthquake sequence.
(4) To collect the data of the earthquake sequence phase and seismic waveform of earthquake above medium level for calculating and analyzing focal mechanism solutions, precise earthquake locating and apparent stress.
(5) To obtain the focal mechanism solutions and precise earthquake locating by inversion method under CRUST1.0 and the appropriate velocity structure selected from the research area and its vicinity.
(6) To collect the seismic geological data relevant to the earthquake area, including active fault data, etc., for providing evidence for the interpretation of focal mechanism solutions and results of precise earthquake locating.
(7) To collect the geological and geophysical research results of the research area and its vicinity for providing evidence for the analysis of the research results of the sequence parameters.

2. Field surveying at site

Generally, there is no related field surveying work at site for the study of evaluating the tendency of earthquake sequence.

3. Analysis and summary indoor

(1) Analysis of seismic activity characteristics in the source area.
(2) The results and plots of the background seismic activity and the distribution of the cluster seismic activity, the temporal evolution of the model parameters, the probability prediction with time and the performance evaluation results.
(3) Results of the probability prediction analysis and evaluation of aftershock occurrence trend.
(4) The geometric structure and kinematic properties of faults are presented by analyzing focal mechanism solutions and results of precise earthquake locating preliminary.
(5) The preliminary results are verified, and the detailed fault structure, stress field of the earthquake sequence are given, combined with the field fault geological survey results in this earthquake and the existing geological and geophysical research results.
(6) Editing of the report.

16.5 Results and Data Archiving

1. Content and form of the results

1) Contents and forms of basic parameters and statistical parameters of seismic sequence.

The results of the basic parameters of seismic sequence are mainly reflected in the analysis reports of emergency consultation, additional consultation and special consultation. After the consultation and scientific survey, the research results are submitted in the form of graphic reports and relevant attachments.

The work results of seismic sequence statistical seismology model are mainly reflected in the characteristics of background seismicity and cluster seismicity in the study area, the time change of model parameters, the prediction of aftershock probability and its efficiency evaluation and are submitted in the form of summary report.

2) The contents and forms of focal mechanism solution and precise location.

Focal mechanism solution parameters of strong earthquakes and maps of focal mechanism solution in the earthquake sequence, the catalogue of precise earthquake location and epicenter distribution, maps of depth distribution, analysis of focal mechanism solution and precise location, the geometric structure, kinematic properties, characteristics of stress field in earthquake area, prediction of subsequent development tendency in the earthquake sequence of earthquake sequence faults, etc., are submitted in the form of summary report.

2. The main content of the earthquake sequence summary report

1) Summary report of the basic parameters and statistical seismic model parameters.

The report includes but is not limited to the following main elements:

(1) Preface.

Including work objectives, work content and workload by time nodes, investigation methods and work deployment, working group personnel list and division of labor.

(2) Tectonic background and characteristic of seismic activity around the focal area.

Including the basic analysis of general geological tectonics and regional seismicity.

(3) Analysis of background seismic activity and cluster seismic activity characteristics.

Including the scope of the study area and the selection of the initial parameters of the model, the fitting of the statistical seismic model of the seismic activity in the source area and the spatial distribution characteristics of the background seismic activity and the cluster seismic activity.

(4) Prediction of the probability aftershock occurrence.

Including the time variation characteristics of the model parameters after the earthquake, the aftershocks occurrence probability prediction of a certain time window after the earthquake and the performance evaluation of the statistical test method.

(5) Issues remained and deficiencies of the study.

To summarize the work of statistical seismology model systematically, compare the research results with the expected goals, and raise questions on the issue remained and deficiencies of the study, then to provide reference for the future work.

2) Main contents of summary report on focal mechanism solution and precise location.

Including five parts: data for study, researching methods, analysis of results, conclusion and tendency prediction of the earthquake sequence.

3. Submission of raw data

1) Analysis of basic seismic sequence parameters and statistical seismology model parameters.

Mainly including: original earthquake catalogue for study, temporal variation of the aftershock sequence parameters, spatial distribution of the background and cluster seismic activity intensity in the study area, prediction results of the occurrence rate and occurrence probability after the earthquake and statistical test results of the probability prediction.

2) Focal mechanism and precise earthquake locating.

The seismic phrase report, waveform data, focal mechanism solution and precise location are submitted. Confidential information shall be archived and managed in accordance with relevant national confidentiality management measures.

Chapter 17
Data Collection and Sharing Services

17.1 Objectives

The seismic scientific expedition involves remote sensing interpretation, seismic geological survey, seismic hazard survey, crustal deformation observation, artificial seismic detection, portable seismic array observation, gravity detection, geomagnetic survey, deep seismic reflection, geochemical observation, source rupture process and seismic tectonic mechanism, earthquake sequence research and so on. All of these will generate a large number of survey results, observation data and research results, which will become the important basic information for understanding the characteristics of the earthquake and the deep seismogenic environment in the source area.

The objective of colleting the data of seismic scientific expedition is to complete the submission of the working achievement of each subject according to the work plan, establish the seismic scientific expedition database and carry out data quality assessment. The database which has passed the quality test serves as an important carrier for the storage, management and sharing services of the seismic scientific expedition data and guarantees the orderly organization, effective management and beneficial application of it.

The objective of the seismic scientific expedition data sharing service is to carry out data analysis, utilization and share rely on the seismic scientific research database, to promote the multidisciplinary integration, support seismic scientific research, technological innovation, public services and other work, and serve the needs of modernization and construction of earthquake prevention and disaster reduction.

The database should be established from the beginning of the seismic scientific expedition. The management of observation data and the sharing of data with the relevant regulations are of great importance to the seismological scientific expedition plan, the promotion of multidisciplinary cross-sectional research and the improvement of innovative capabilities.

Institute of Earthquake Forecasting, China Earthquake Administration,
A Guidebook to Earthquake Scientific Investigation,
https://doi.org/10.1007/978-981-96-0418-0_17

17.2 Contents

1. Establishment of a management method for the aggregation of the seismic scientific expedition data

Under the guidance of the competent department, the leading unit should organize to formulate the management method of the seismic scientific expedition data submission, specifying the objectives, principles, requirements, contents, forms, quality control, management and other specific requirements of data submission, and may formulate the implementation rules of data submission according to the contents and objects of different scientific research. Establish and improve a system consisting of seismic professional standards, guidelines, technical specifications and data management method for the seismic scientific expedition to standardize and strengthen data transfer management and ensure the successfully integration of seismic scientific research results.

The submission of data results should be set as an important part of for the seismic scientific expedition plan. The contents and forms of data submission should be clearly defined in each subject, and the submission of data and the passing of quality assessment should be made as a pre-condition for the summary acceptance of the tasks.

2. Design and establishment of a database for the seismic scientific expedition

The leading unit of seismic scientific expedition should organize to formulate the database specification including of designing the seismic science research data model and database structure based on a GIS platform, establishing metadata information tables and a common data dictionary, and developing and building a unified data entry work platform.

According to the requirements of seismic standards and related technical specifications, the database schemes of the subjects such as seismic geological survey, earthquake disaster remote sensing comparison survey and disaster acquisition, crustal deformation observation, geochemical observation, source rupture process and seismic tectonic mechanism, artificial seismic detection, gravity profile detection, portable gravity detection, geomagnetic survey, portable seismic array observation, mobile geomagnetic observation, strong ground motion and engineering earthquake damage investigation and earthquake sequence research should be designed respectively with metadata tables. The database tables of each subject are integrated into the general database of seismic science research, and the general database completes the integration of basic database tables such as coding rules, cross data, associated data, statistical data and task information.

The database structure of each topic in the seismic scientific expedition database is compatible with existing databases, such as the national active fault detection database operated by the Seismic Active Fault Detection Data Center and the station array database operated by the China Seismic Scientific Detection Station Data Center, making full use of the existing work base and facilitating data exchange and shared applications.

3. Development of a work program for data submission

As the seismic scientific expedition units are involved in a wide range of work and produce a wealth of results, a detailed work plan is needed to support the scientific and effective completion of the data transfer. As an important part of the overall plan, the data submission program should include a detailed list of data submission according to the content and characteristics of each topic, and specify the specific requirements on the content, form, quality, contact person and deadline for the submission of archival data, electronic files and databases.

4. Organization and management of the data transfer

Under the guidance of the competent department, the leading unit which is in charge of the database construction and data submission management, is responsible for formulating the data submission management methods and implementation rules, designing and building the database template and working platform and organizing data submission, quality assessment, storage management and sharing services.

Establish data archiving and sharing collaboration mechanisms with relevant units under the Bureau, in order to cooperate in database construction, template design, data archiving and quality testing, make full use of the existing work base, improve efficiency and promote data sharing and exchange.

On the basis of the existing hardware and software environment, a hardware and software system suitable for the requirements of this earthquake will be established by reasonably increasing hardware equipment and developing and compiling relevant programs, etc. The data security will be enhanced by combining data backup, permanent online data storage, regular integrity and consistency checks.

For example, by relying on the Seismic Fault Detection Data Center, the real-time observation data of faults and surface rupture zones obtained during the implementation of scientific expedition are integrated with the data of faults and folds to provide seismic tectonic determination data services. By relying on the China Earthquake Scientific Exploration Data Center, the real-time observation data of the scientific research arrays are fused with the real-time transmission data of the Himalayan arrays, Qiaojia arrays and Xichang arrays to provide waveform and event data services. The observation data of the arrays collected by mobile media are archived and stored in the database to provide long-term scientific research arrays data services. The observation data of the portable observation, profile detection and other data are stored in a metadata and index database, and the observation data are stored in a file system to provide services to the public. For confidential data, they are stored in a confidential room after collation and are regularly checked for correctness.

17.3 Methods and Processes

Figure 17.1 shows the general flow of the data aggregation and service work.

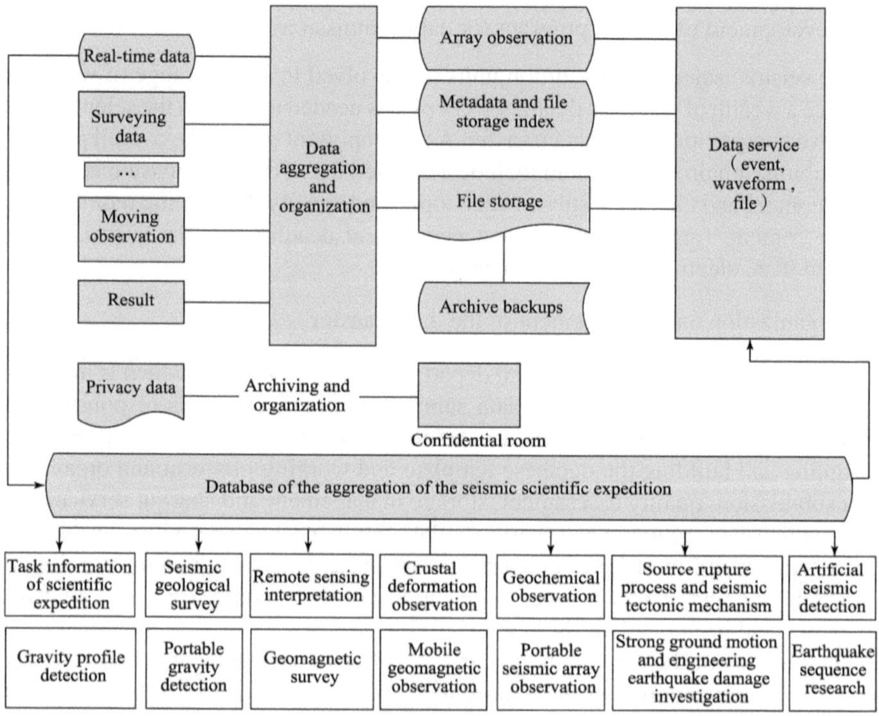

Fig. 17.1 Workflow diagram for data submission and services

17.4 Results and Data Archiving

1. Content and form

Each subject should prepare the corresponding expedition report and submit the raw data such as field observation data, test and laboratory data, drawings, photos and images as annexes to the unit undertaking of seismic scientific expedition or the headquarters.

Content and form of data archiving is shown in Fig. 17.2, using the seismic geology section as an example.

2. Main elements of the investigation report

The report should include and not be limited to the following main elements:

(1) Preface.

Including research objectives, content and so on for each subject.

(2) Data processing process.

The process of collating data and constructing database for each subject.

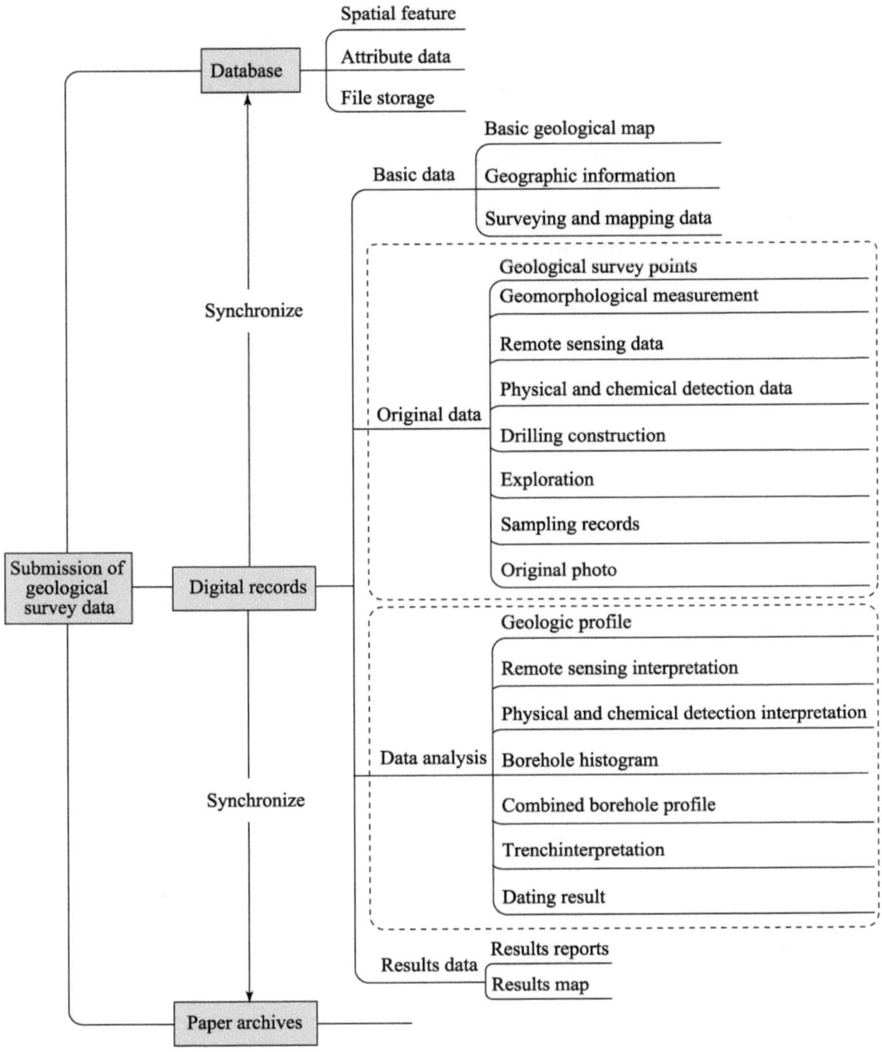

Fig. 17.2 Workflow diagram submission of geological survey data

(3) Summary of results submission.

A comprehensive summary of the results of the scientific expedition research was carried out. With comparing with the expected objectives, the suggesting problems and shortcomings to provide a reference for future submission and sharing of scientific research data should be proposed.

3. Raw data submission

Raw data includes, but is not limited to the following.

(1) Basic geological, mapping and surveying information covered by each subject group.
(2) Various raw survey (mapping and surveying) data acquired during field work.
(3) Data from testing and laboratory analyses.
(4) Various types of field work photographs (collections), video materials and their textual descriptions, etc.
(5) Confidential information is archived and managed in accordance with the relevant national confidentiality management methods.

Appendix

During the process of compiling the Guidebook, some technical standards, state and professional norms are referred. Relevant laws and regulations that may be involved in the seismic investigation, related documents of China Earthquake Administration (and its subordinate departments), and some technical standards and norms not directly cited in the text are listed here so that readers can refer themselves when needed.

[1] *Safety specifications of blasting* (GB/6722-2011).
[2] *Safety regulations for seismic exploration blasting practices* (GB 12950-1991).
[3] *Post-earthquake field works-part 3: Code for field survey* (GB/T 18208.3-2011).
[4] *Post-earthquake field works-part 4: Assessment of direct loss* (GB/T 18208.4-2011).
[5] *Standard for labor protective supplies and emergency lifesaving supplies of geological exploration* (AQ 2049-2013).
[6] *Safety regulations for geological prospecting operation* (AQ 2004-2005).
[7] *Surveying and prospecting of active fault* (GB/T 36072-2018).
[8] *Mapping of active fault in 1:50,000* (DB/T 53-2013).
[9] *Code of 1:50,000 active fault mapping database* (DB/T 65-2016).
[10] *Active fault survey—Compilation of 1:250,000 seismic structure map* (DB/T 73-2018).
[11] *Active fault survey—Measurement of faulted landform* (DB/T 71-2018).
[12] *Active fault survey—Paleoseismic trenching* (DB/T 81-2020).
[13] *Active fault survey—Inspection of database* (DB/T 83-2020).
[14] *Active fault survey—Graphic symbols* (DB/T 72-2018).
[15] *Active fault survey—Remote sensing investigation* (DB/T 69-2017).
[16] *Active fault survey—Field geological investigation* (DB/T 82-2020).
[17] *Classification of earthquake damage to buildings and special structures* (GB/T 24335-2009).

Institute of Earthquake Forecasting, China Earthquake Administration,
A Guidebook to Earthquake Scientific Investigation,
https://doi.org/10.1007/978-981-96-0418-0

[18] *Technical specifications for land seismic exploration data processing* (SY/T 5332-2011).
[19] *Technical code for land 2D seismic exploration data acquisition* (SY/T 5330-2003).
[20] *Technical regulations for strong motion observation* (DB/T 64-2016).
[21] *Classification of earthquake damage to lifeline engineering* (GB/T 24336-2009).
[22] *Digital strong motion accelerograph* (DB/T 10-2016).
[23] *Technical specifications for acquisition of natural source electromagnetic sounding* (SY/T 5820-2020).
[24] *Chinese seismic intensity scale* (GB/T 17742-2020).
[25] *The technical specification for gravity survey* (DZ/T 0004-2015).
[26] *Law of the People's Republic of China on Work Safety.*
[27] *Law of the People's Republic of China on Guarding State Secrets.*
[28] *Law of the People's Republic of China on Road Traffic Safety.*
[29] *Law of the People's Republic of China on Protecting Against and Mitigating Earthquake Disasters.*
[30] *Law of the People's Republic of China on State Security.*
[31] *Labor Law of the People's Republic of China.*
[32] *Outline of Urban Active Fault Survey.*
[33] *Administrative Measures for Seismic Scientific Data Sharing.*
[34] *Regulations on Management of Post-earthquake Field Works.*
[35] *Emergency Response Plan Scenario for Destructive Earthquake.*
[36] *Administrative Measures for Active Fault Detection.*
[37] *Emergency Regulations for Destructive Earthquake.*
[38] *Response Grade of Earthquake Emergency Service of China Earthquake Administration (Provisional).*
[39] *Earthquake Emergency Plan of China Earthquake Administration.*
[40] *Work Plan for Seismic Scientific Investigation (Trial).*
[41] Engineering Research Center for Crustal Movement Monitoring. Crustal Movement Monitor [M]. Beijing: China Environmental Science Press, 2013.
[42] China Earthquake Administration. Code for Seismic Gravimetry [M]. Beijing: Seismological Press, 1997.
[43] Hach Company. Practical Manual on Water Quality Analysis (Second Edition) [M]. Beijing: Chemical Industry Press, 2016.
[44] Dafeng Huang, Xiuquan Luo, Xibin Li, et al. Isotope Mass Spectrometry Technology and Application [M]. Beijing: Chemical Industry Press, 2006.
[45] Shoude Lu. "Earthquake Field Work" National Standard Publicity Textbook (Part 1) [M]. Beijing: Seismological Press, 2002.
[46] Shihua Qi. Geothermal Geochemical Survey [M]. Beijing: Science Press, 2021.
[47] Hong Shi, Xinzhong Sun, Jianhua Fan, et al. Methods and Techniques for Water Quality Analysis [M]. Beijing: Seismological Press, 2001.
[48] Yuyu Xia, Yan Zhu, Jie Li. Chemical Laboratory Handbook (Third Edition) [M]. Beijing: Chemical Industry Press, 2015.

[49] China Earthquake Administration. Outline and Technical Guide for Earthquake Field Work [M]. Beijing: Seismological Press, 1998.
[50] China Geological Survey. Handbook of Hydrogeology (Second Edition) [M]. Beijing: Seismological Press, 2012.

MIX
Papier aus verantwortungsvollen Quellen
Paper from responsible sources
FSC® C105338

If you have any concerns about our products,
you can contact us on
ProductSafety@springernature.com

In case Publisher is established outside the EU,
the EU authorized representative is:
Springer Nature Customer Service Center GmbH
Europaplatz 3, 69115 Heidelberg, Germany

Printed by Libri Plureos GmbH
in Hamburg, Germany